STUDENT UNIT GUIDE

AS Geography
UNIT 2

Edexcel

Specification A

Unit 2: Human Environments

Peter Goddard & Nigel Yates

AS Geography

Philip Allan Updates
Market Place
Deddington
Oxfordshire
OX15 0SE

tel: 01869 338652
fax: 01869 337590
e-mail: sales@philipallan.co.uk
www.philipallan.co.uk

© Philip Allan Updates 2002

ISBN 0 86003 727 4

All rights reserved; no part of this publication may be reproduced, stored in a retrieval system, or transmitted, in any form or by any means, electronic, mechanical, photocopying, recording or otherwise without either the prior written permission of Philip Allan Updates or a licence permitting restricted copying in the United Kingdom issued by the Copyright Licensing Agency Ltd, 90 Tottenham Court Road, London W1P 9HE.

This Guide has been written specifically to support students preparing for the Edexcel Specification A AS Geography Unit 2 examination. The content has been neither approved nor endorsed by Edexcel and remains the sole responsibility of the authors.

Printed by Raithby, Lawrence & Co. Ltd, Leicester

Edexcel (A) Unit 2

Contents

Introduction
About this guide 4
AS Geography 4
Geographical resources, skills and techniques 5
Revision advice 7
Investigative skills 8
Examination skills 9

■ ■ ■

Content Guidance
About this section 12

Population characteristics
Populations vary in distribution, density and structure 13
Populations change over time 21
Population changes have socio-economic and environmental implications 28

Settlement patterns
The size and location of settlements vary, and distinct patterns can be
   identified 32
Settlements vary in their internal structure 40
Settlements change over time 46

Population movements
Population movements can be classified 50
There is a variety of causes and constraints affecting people's decisions to
   migrate permanently 57

■ ■ ■

Questions and Answers
About this section 64
Q1 Population characteristics (I) 65
Q2 Population characteristics (II) 69
Q3 Settlement patterns (I) 73
Q4 Settlement patterns (II) 77
Q5 Population movements (I) 80
Q6 Population movements (II) 83

Introduction

About this guide

This guide is for students following the Edexcel Specification A AS Geography course. It aims to guide you through Unit Test 2, which examines the content of **Unit 2: Human Environments**.

This guide will clarify:
- the content of the unit so that you know and understand what you have to learn
- the nature of the unit test
- the geographical skills and techniques that you will need to know for the assessment
- the standards you will need to reach to achieve a particular grade
- the examination techniques you will require to improve your performance and maximise your achievement

This **Introduction** describes the structure of AS Geography and outlines the aims and method of examining Unit 2. It then provides advice on learning and revision techniques before explaining some of the key command words used in examination papers. There is also advice concerning general and specific geographical skills.

The **Content Guidance** section summarises the essential information of Unit 2. It is designed to make you aware of the material that has to be covered and learnt. In particular, the meaning of key terms is made clear.

The **Question and Answer** section provides sample questions and candidate responses at C-grade level and A-grade level. Each answer is followed by a detailed examiner's response. It is suggested that you read through the relevant topic area in the Content Guidance section before attempting a question from the Question and Answer section, and only read the specimen answers and examiner's comments after you have tackled the question yourself.

AS Geography

AS is designed to be an intermediate standard between GCSE and A-level. While it is difficult to specify exactly what that standard is, it recognises that students will only have followed the course for one year.

After you have completed the AS course, you may decide to stop studying the subject. Alternatively, you can continue with the A2 course, which has different units and is designed to be more demanding. This will ultimately enable you to combine your AS results with your A2 results and to receive an A-level grade.

Edexcel (A) Unit 2

Scheme of assessment

Unit 2: Human Environments is one of three units that make up the AS specification. It is assessed by a written paper which is marked out of 60 marks and worth 90 uniform marks (i.e. 30%) of the 300 marks that make up the whole assessment.

The AS marks make up 50% of the total A-level assessment. The three A2 units make up the other 50%.

Unit	Unit exam length	Max mark	Max uniform mark	AS weighting
1: Physical Environments	1 hour 15 minutes	60	90	30%
2: Human Environments	1 hour 15 minutes	60	90	30%
3: Applied Geographical Skills	1 hour 30 minutes or Personal Enquiry	60	120	40%

Unit 2

The specification content of Unit 2 comprises three sections:
- Population characteristics (at global, national, regional and local scales)
- Settlement patterns (at national, regional and local scales)
- Population movements (at international, national, regional and local scales)

Unit 2 aims to help you:
- acquire and apply knowledge and understanding of processes in human geography
- understand the influence, interactions and outcomes these processes have on people, and how they change over space and time
- apply this knowledge and understanding at a variety of scales
- develop an understanding of the interrelationships between people and their environments, and how places and issues change over time
- learn and apply geographical skills, understand that geography is dynamic, and that change necessitates a response
- reflect the importance of people's values and attitudes in relation to geographical issues and questions

Geographical resources, skills and techniques

You are expected to use a variety of geographical resources, skills and techniques (RSTs) in the various unit tests. Listed below are the RSTs you will need for the AS units, followed by some specific to Unit 2.

General geographical skills needed for all AS units

- Topographical maps (especially OS 1:25 000 and 1:50 000)
- Land-use maps
- Choropleth and isopleth maps
- Aerial and satellite photographs
- Annotated sketches from the field or from photographs
- Annotated sketch maps
- Line graphs and cumulative line graphs
- Bar charts and histograms
- Pie graphs and divided bars
- Scatter graphs, best-fit lines and/or curves
- Triangular graphs
- Flow lines
- Sketch sections, cross-sections and long sections
- Proportional symbols
- Mean, mode, median
- Quartiles and inter-quartile ranges
- Spearman rank correlation coefficient
- Chi-squared test

Specific geographical skills needed for Unit 2

Population characteristics
- Population census data — used by national governments for planning and policy making
- Population pyramids — show amounts by groups (otherwise known as age versus sex pyramids)
- Dependency ratios — attempt to compare those who are economically active and those who are either too old or too young to work
- Lorenz curves — show inequalities in distributions (an even distribution is located around a 45° line; a varied distribution shows distributions away from this line)

Settlement patterns
- Rank-size rule — shows the relationship between the sizes of settlements in a country
- Break-point analysis — used to draw boundary lines showing the limits of trading areas
- Nearest neighbour analysis — a statistical test used to describe settlement patterns
- Logarithmic graphs — used to show data that might not fit on an arithmetic scale; or where there are data that might be concentrated at different ends of a range
- Rates of change, not amounts of change, assessed on log paper

Population movements
- Gravity models — used to predict the degree of interaction between two places

Edexcel (A) Unit 2

Revision advice

Command words

You need to understand the command words used in the examination to make the most effective use of your time. The following command words are typically used in Unit 2. The words are set out in an approximate order of difficulty and a brief description is offered for each.

- **Describe:** give details of the appearance and characteristics
- **Define:** give the meaning of
- **State:** give a very brief, possibly one-word, answer
- **Suggest:** offer possible reasons that logically/reasonably might be appropriate
- **Illustrate:** show with the use of an example(s) or a case study
- **How:** identify the process or mechanism
- **Outline:** provide general principles or main features without great detail
- **Why:** explain
- **Explain:** give reasons for

General learning tips

Most of you will be taking the AS examination at the end of or during a 1-year course. There is no surplus time available for teaching the subject content, so you must ensure that from the start of the course, you establish good working practices to make the most of the time available.

It is important that you do not fall behind with work during the year. New material will be taught each week, so if you are unavoidably absent because of illness, for example, do make sure you are able to make up the missed work as quickly as possible.

Read widely from a variety of sources. Television programmes are also relevant. The information you gather will enable you to develop a number of case studies for use in your examination answers.

If you keep on top of the work, your revision programme will be more relevant and straightforward in the lead-up to exams.

General revision tips

- Having selected a topic for revision, read and learn the material you have for this topic, for example notes, handouts and worksheets.
- Refer to your textbooks and to this publication. You might also find Raw, M. (2000) *AS/A-level Geography Exam Revision Notes* (Philip Allan Updates) a useful guide.
- Learn the relevant case studies. For AS, you probably need no more than two for each section/sub-section. These should be at different scales to meet the examination requirements.

- Practise sample questions, keeping to the appropriate timings. Use the Question and Answer section of this guide, taking care not to look at the sample answers and examiner comments until you have attempted the questions. There are other specimen questions available, so consult your teacher/lecturer for advice.
- Apply your knowledge and understanding when practising, so that your answers reflect the demands of the question.
- Allow yourself adequate time for revision. Little and often is usually better than concentrated pressure at the last minute.

Revising for Unit 2

The three sections of Unit 2 are sub-divided as follows:

Sections	Sub-sections
Population characteristics	• Populations vary in distribution, density and structure • Populations change over time • Population changes have socio-economic and environmental implications
Settlement patterns	• The size and location of settlements vary, and distinct patterns can be identified • Settlements vary in their internal structure • Settlements change over time
Population movements	• Population movements can be classified • There are a variety of causes and constraints affecting people's decisions to migrate permanently • Migrations have an effect on the areas people have left, and the areas they move to

Revision can be more easily structured by taking the sub-elements and focusing on these. Note that it is better to revise the sub-elements in the order in which they appear; otherwise some points might not make sense.

Investigative skills

Identification of geographical questions and issues

Unit 2 provides a wide range of opportunities for investigative study. Settlement studies are the main focus of much fieldwork, though population and industry also provide opportunities.

Selection of relevant primary and secondary data and an assessment of their validity

The popularity of settlement-related fieldwork means that appropriate data can be collected. The nature of the hypotheses to be tested will define the type of data collected, for example land use, land values, questionnaires, and so on. Secondary data on settlements are available from a number of sources, including the Office of Population Censuses and Surveys, and local authorities. The internet is also a valuable means of accessing such data.

You also need to understand the reliability of the methods of data collection employed and/or the data collected.

Processing, presentation, analysis and interpretation of the evidence collected

These skills are developed as the course progresses and are not confined to data collected in the field. They could be assessed in a general sense in questions in Units 1 and 2 and will certainly be assessed in the applied geographical skills paper, Unit 3b. Writing up fieldwork and analysis of data downloaded from the Internet will develop such skills. In terms of population settlement, retailing and economic activity, a comparison of the results, interpreted in relation to theory, will fulfil the requirements of this section.

Ability to draw conclusions and show an awareness of their validity

This skill is again not confined to data collected in the field. It could be assessed in the general sense when responding to questions in Units 1 and 2, and will certainly be assessed in Unit 3b. A conclusion to, and an evaluation of, the success of any geographical fieldwork exercise would be appropriate here.

Awareness of risks when undertaking fieldwork

This is essential. For example, it is not advisable to undertake fieldwork on your own. Assistance from others is needed to collect the data and to ensure your safety in any geographical fieldwork situation.

Investigative skills are best developed by a programme or fieldwork undertaken in the AS year. Preparation for work in the field, the collection of data and their interpretation and evaluation are demonstrated clearly by writing up the fieldwork in the format suggested by the board. Your teachers will be able to advise you on this.

Examination skills

There are six structured data–response type questions, two from each of the three elements. You have to answer one question from each section.

Each question is marked out of 20, making a total of 60 marks for the whole paper. You have 25 minutes to complete each question; this includes 5 minutes or so reading and thinking time. The spacing left after questions for your response allows two lines per mark. It is important to allocate your time according to the line and mark allocations.

Question structure

Parts (a) and (b)

Each question will include an item of stimulus material, such as a map (OS 1:50 000 or 1:25 000), a table of data, a graphical representation or a diagram. Part (a) of the question will be related directly to this material, with candidates being required to manipulate or interpret the information provided. Later parts of the question may depart from the original focus and refer to other parts or strands from the same section of the unit. This ensures a greater coverage of the specification. Parts (a) and (b) are generally awarded marks in the range 1–4. Occasionally, question parts with a high mark award are levels-marked (see page 64).

Part (c)

The final part of questions in this unit require some extended writing and you will incorporate your own examples or case studies within your answer. Most marks (6–8) are reserved for part (c) of each question, so ensure that you have put 7 or 8 minutes aside to answer it!

Should your response exceed the line allocation, try to use the space adjacent to each question for your answer overrun. If you need extra paper, make it very clear to which question your answer belongs. There is no penalty for overrunning!

Part (c) is always levels-marked (see page 64), because examiners have to establish clear and appropriate differentiation between basic and higher-level responses. These levels can be revised in the light of candidates' responses. You are then being 'assessed on your ability to organise and present information, ideas, descriptions and arguments clearly and logically, taking account of your use of grammar, punctuation and spelling' (quoted from the cover of the Unit Test 2 question booklet).

Content Guidance

There are three sections in the specification content for Unit 2:

(1) Population characteristics

(2) Settlement patterns

(3) Population movements

In this Content Guidance, each of the three elements will be considered in terms of:
- the key concepts involved
- the content required
- the use of examples and case studies

This should make it clear to you what you need to know and understand.

It should be noted that the synoptic links in each element are not dealt with in this guide. Although these may be taught as part of the AS course, they are not assessed until A2.

Edexcel (A) Unit 2

Population characteristics

Populations vary in distribution, density and structure

Physical and human factors affecting distribution of population, including a study of population distribution in the UK

- *You need a good overview of the main features of global population distribution.*
- *You should be able to explain variations in distribution and density, allowing physical factors pre-eminence at this scale, but avoiding environmental determinism.*
- *You should know the broad pattern of the UK distribution and understand the importance of patterns that were established by nineteenth and early twentieth century processes of industrial growth and development. You need to recognise the modification of these patterns during the late twentieth century by fundamental changes in employment, a revolution in transport and a significant government involvement in planning and control.*

The **population distribution** of a country refers to the way in which people are spread out across the Earth's surface. There are several **population density** measures; generally, these indicate the number of people per unit area. In effect, they show the number of people who would be living in each square kilometre if the population was evenly distributed over the land.

Both the distribution and density of population vary locally, regionally, nationally and internationally.

The variables that control population distribution can be very broadly classified as either **physical** or **human**.

Physical factors

Climate — this is the dominant factor at a global scale and important at a national level. The dominant elements are:

- temperature, which limits the length of the thermal growing season and thus limits agricultural production
- rainfall and water availability. There is no upper limit here, but annual rainfall figures below 250 mm per annum inhibit permanent settlement. They do not actually prevent it where other water sources are available, as in the Nile valley, but they place very severe limits on agriculture.

Soil — this becomes more significant at the national and, especially, the local level. Soils of low fertility are generally closely related to particular climate types. Thus, desert soils are thin and lack nutrients as a consequence of limited organic matter input and low rates of chemical weathering. This relationship is quite complex in a vast region like Amazonia, where the soils become infertile if the natural vegetation is removed (as is the case with most agricultural systems). It is not that the soil is inherently infertile, but it is its inability to support intensive agriculture which limits population.

Relief — slope angle operates on a local scale. Steep slopes can be limiting both because of their influence on soil depth and soil fertility and as a result of the practical problem of working on them. Don't confuse relief with altitude.

Altitude — population density decreases with increasing altitude in most parts of the world. There are some significant anomalies to this 'rule' close to the equator (e.g. Ecuador and Kenya). The impact of high altitude is a reduction in the thermal growing season and significant increases in precipitation. Where the coastal climate is tropical (e.g. Ecuador and Kenya) this can be beneficial for human habitation.

Human factors

Economic development — the point here is that industrial development frees people from a dependence on agriculture. Large concentrations of population can be supported by small areas of land producing goods of sufficient value to exchange for the food that they are unable to grow themselves in urban areas. (The industrial revolution in Britain was associated with a sharp increase in population.) Trade also frees countries from a dependence on their local physical conditions.

Historical factors — some parts of the world are relatively lightly populated because humans have not been in occupation as long as in other areas. The last continent we reached as a species was South America; this provides some explanation for its relatively low overall population density. The development of very high densities in parts of Asia is a result of the development of irrigation systems and the wet-rice agriculture which these systems have supported. In Java, for example, wet-rice agriculture supports a population of more than 1000 per km^2. This is a particularly instructive example given that it has more or less the same climate as Amazonia. It therefore appears to undermine the often quoted 'too hot, too humid and too many diseases' explanation for the Amazon basin's very low population density. However, you should remember that Java has immensely rich volcanic soils and that wet-rice agriculture can, and has been, maintained for many millennia, without interruption, because the irrigation process naturally fertilises the soil, bringing silts in from the river's load. Rice is not a native species of South America but is becoming increasingly important.

Government policies — governments can move people around both directly (in some societies) and indirectly (through planning and economic management). Decisions are made about where people can and cannot live and there are very many examples of governments forcing people to move, as you will find out in the third section of this unit.

Other factors affecting population distribution include water supply, diseases and pests, resources and communications. In LEDCs, there is a strong link between the physical, climatic and hydrological environment (e.g. northern India).

As the global population has doubled over the past 40 years, the shifts in geographical distribution of that population have been remarkable. In 1960, 2.1 billion of the world's 3 billion people lived in LEDCs (70% of the global population). By late 1999, the LEDCs had grown to 4.8 billion (80%); by 2025, 98% of the projected growth of the world population will occur in these regions.

Edexcel (A) Unit 2

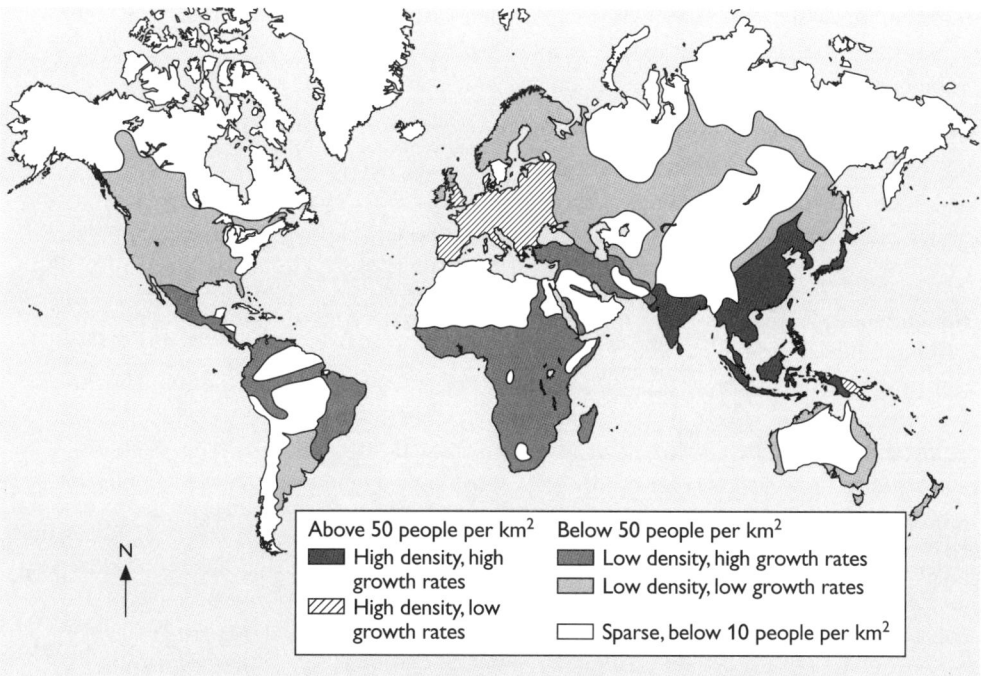

Figure 1 World distribution and density of population

Africa, with an average fertility rate over the entire period exceeding five children per woman, has grown the fastest among these regions. There are almost three times as many Africans alive today (767 million) as there were in 1960. Asia, by far the most populous region, has more than doubled in size (to over 3.6 billion), as has Latin America and the Caribbean. By contrast, the population of North America has grown by only 50%; Europe's has increased by only 20% and is now roughly stable.

Africa's share of the global population is projected to rise to 20% in 2050 (from only 9% in 1960), while Europe's share is projected to decline from 20% to 7% over that same period. In 1960, Africa had less than half the population of Europe; in 2050 it may have as much as three times as many people.

The altered balance of population distribution among regions does not pose a problem, so long as development progresses everywhere and population growth is balanced by the development of social and economic capacity. The challenge remains to create conditions that will enable countries in all regions to adopt policies and strategies that foster equitable development.

The UK's population distribution
The UK's population density of 250 people per km² is relatively high, but it is a very uneven distribution. Some areas are very lightly populated — the Highlands of Scotland have 10–20 people per km². Most of the lightly populated areas of the UK have relatively severe physical limitations which in pre-modern times made them difficult

for human settlement. Although we are no longer constrained by such physical factors, many of these areas have, paradoxically, lost population.

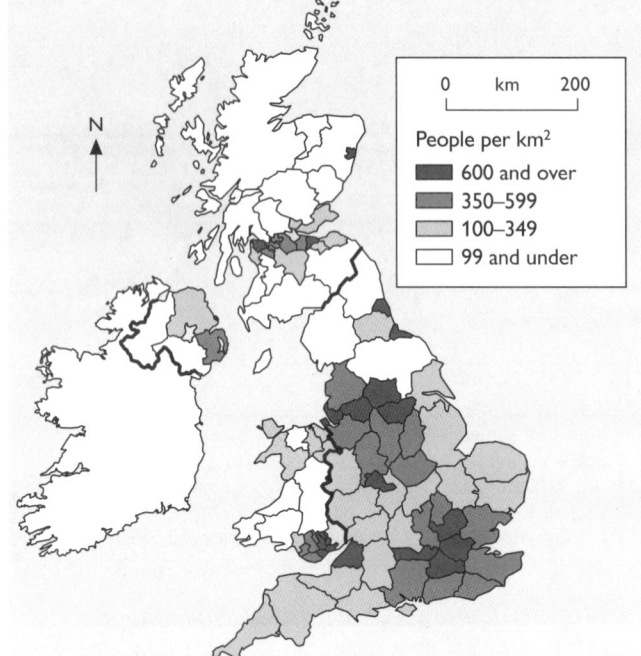

Figure 2 Distribution and density of population in the UK

Some areas are very highly populated, for example London and the southeast average 1500 people per sq km. This distribution is polarised at a local scale — Salisbury Plain has only 1–5 people per km^2, while parts of the London Borough of Camden hold over 30 000 per km^2.

Factors accounting for variations in the UK distribution

Physical restaints on agriculture include:
- rainfall (more than 1500 mm makes agriculture difficult)
- the thermal growing season, which shows a rapid decline with altitude (6–7 months in the Pennines and north Wales)
- soils — these can be thin and acidic on the moorlands of much of highland Britain, while some lowland soils can be impoverished and sandy (e.g. Breckland in East Anglia)

Economic factors can counteract these physical constraints. For example, South Wales has a poor climate and soils, but the presence of coal resulted in a concentration of people and the creation of vital ports for trade. The same applied to other coal-producing areas.

The role of government has become more important. 'New Towns' were a post-1945 attempt to relieve the pressure on London and its surrounding green-belt by building new settlements on what we would now call 'greenfield' sites. Governments now use planning to influence the growth and development of settlements. For instance, the distribution of population in many rural areas depends on which villages are designated as 'key settlements' and permitted to expand should there be a need, while the rest are conserved.

Changes in the distribution

Figure 2 is a snapshot of the UK population and does not illustrate its dynamic nature. The broad distribution still reflects the industrial pattern established over 150 years ago — a dense population in the southeast and midlands, stretching up to Lancashire and Yorkshire. Outside this band, concentrations in the conurbations such as Newcastle and Glasgow grew rapidly in the period between the 1820s and the 1950s.

The old conurbations have lost population (London dropped from 9 million in 1961 to 7 million in 1991), although in most cases, particularly London, this has not in any way reduced their significance in the context of the regional and national economy.

Changes in distribution can be explained in terms of **centrifugal** and **centripetal** forces. For example, counterurbanisation is a centrifugal force — people disperse as a result of better transport facilities, cheaper travel and industry relocating to greenfield sites. Centripetal forces include the growth of industry and therefore the need for the labour force to live at specific points/areas (industrial **agglomeration**).

Inertia is the tendency for the pattern of distribution still to reflect old forces and processes. People still migrate and distributions change, but relatively slowly, as in South Wales where the population decline is not as fast as one would expect given the heavy and rapid industrial loss.

'New' areas of rising density include coastal areas preferred for retirement (e.g. South Devon and Dorset) and the counties around large cities (e.g. Buckinghamshire), both reflecting counterurbanisation trends. Note that the coastal areas are not densely populated compared with many other areas, but they have experienced growth, thus illustrating a changing distribution.

Reasons for variations in density in rural and urban areas

- *You should study at least two rural areas with contrasting population densities.*
- *You should study at least one urban area so that you can both describe and explain the variations in population density.*
- *You should appreciate some of the difficulties involved in defining rural areas and urban areas.*
- *Example areas should be chosen with a view to demonstrating contrasts. Thus an area heavily affected by a nearby urban area such as Buckinghamshire or Cheshire might usefully be compared with a more remote area such as Central Wales or Cornwall.*

Rural areas

The definition of a rural area is not as straightforward as it might seem. There is no generally agreed rule allowing us to distinguish between urban areas and rural areas. In some parts of the world, especially in the USA, they can visibly merge into one another as one travels for miles through a sort of suburbanised countryside.

Rural areas used to be dominated by agriculture. Settlements were small and dominated by people working on the land or in some way dependent upon the land or other primary resources (e.g. fishing villages). In MEDCs, very few villages in a country like the UK have more than 15% of the population dependent upon agriculture, while in some parts of southern England, agricultural workers now live in local towns, commuting to the countryside to work because they cannot compete with commuters living in the rural areas who, generally, have much higher salaries.

Thus, the economic distinction between urban and rural areas has broken down and we are left with size of settlement, which is arbitrary and rather unsatisfactory.

Urban areas

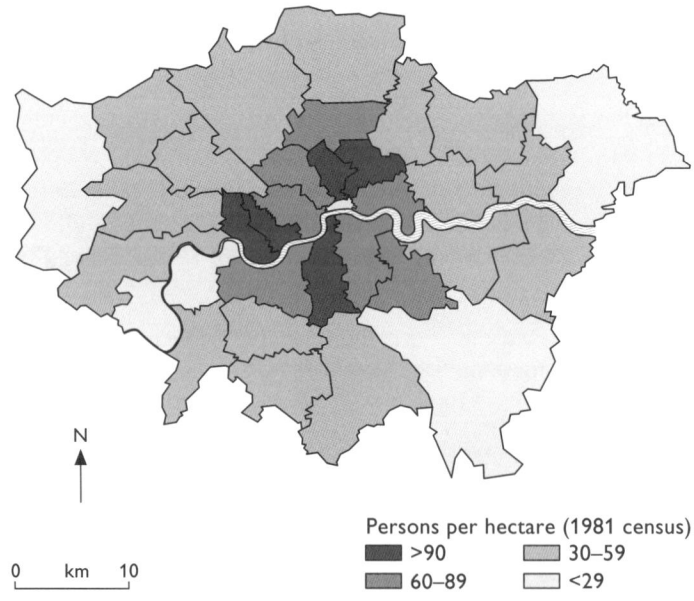

Figure 3 The density of population in London

The following urban characteristics will also give you a check list to apply to your own case-study:
- In many MEDCs, the older, inner suburbs have higher population densities than the twentieth-century suburb areas.
- Some twentieth-century developments involved large-scale municipal housing schemes on the outskirts of the built-up area. These often had higher densities.
- Many CBDs have a very low permanent population, but daytime populations are very high. This is not always the case, as exemplified by central Paris where some local districts have high permanent populations.
- Inner-city redevelopment schemes such as Cardiff Bay and London Docklands can increase urban densities in the inner-city areas.
- Some areas in cities have remained free from any development and thus have very low densities (e.g. Hyde Park in London and Central Park in New York). These

protected areas are entirely a consequence of planning regulations taking over from the protection of their private owners. There are still a few of these privately owned areas about: Buckingham Palace is an obvious example.

The Lorenz curve

This is a technique that attempts to show inequalities in observed distributions. On its straight diagonal axis, one might plot a perfect distribution of population. When actual data are applied to the graph, the curve will show the level at which population is concentrated within various areas in the country under investigation. If, for instance, there is a large area between the diagonal and the curve, there will tend to be a greater concentration of population.

Characteristics of structure and population pyramids of both LEDCs and MEDCs, at national scale and local scale — rural and urban areas

- *Use at least two contrasting countries to illustrate differences in population structure. Contrasts can be drawn between societies with very different economic and social structures. Differences within those societies can be illustrated, for instance between core (usually urban) areas and peripheral (usually rural) areas.*
- *Be aware that population structure is dynamic and that pyramids contain features that are explicable only by reference to policies or processes that operated in the past, for example the effect of the Second World War on the Russian age–sex structure.*
- *Remember that pyramids alone tell you nothing about what has happened or what is going to happen. If one cohort (each slab on a pyramid is known as a cohort) is smaller than the one below it, this could be explained by:*
  — *a decline in birth rate*
  — *an increase in death rate*
  — *migration*
  — *any combination of the above*

Population structure

One of the most important factors influencing changes in population is the number and proportion of people in various sex/age group cohorts. A division of the population into age and gender is fundamental to the population structure. Other potential divisions include religion, ethnic group, language and educational attainment. The divisions by gender and proportions of each sex are drawn onto population pyramids.

The main purpose of drawing pyramids is to show contrasts between populations. For instance, LEDCs tend to have many more people in the youthful pyramid base; this contrasts markedly with the MEDCs, which tend to have many more people in the top half of the pyramid and lower numbers of younger people, giving the pyramids a rather top-heavy appearance.

Once patterns have been established, it is possible to use the information for planning purposes. For instance, the requirements for school places can be assessed, cohorts of OAPs can be provided for and so on. Mobility can be assessed, and the effects of individual 'events' can be chronicled — wars, famines and the like. Population pyramids are able to record and summarise much to do with the demography of a country.

Sweden provides a good MEDC example (Figure 4).

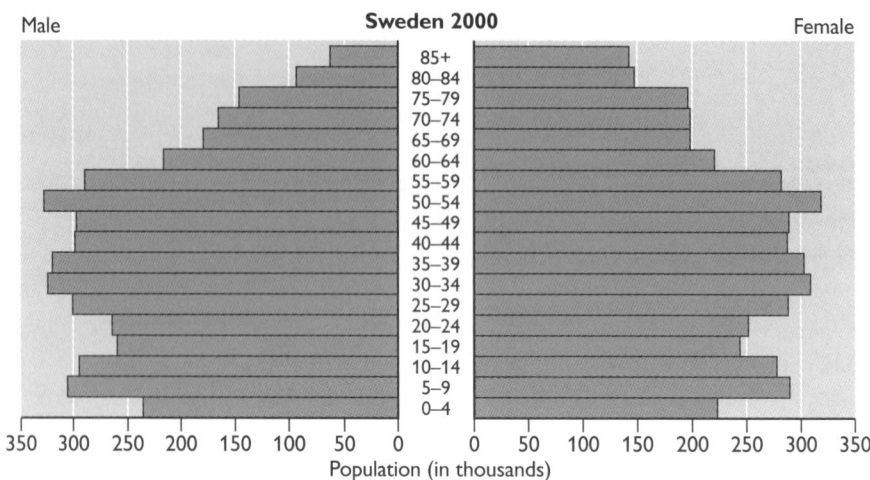

Figure 4 Sweden's population structure

This demonstrates the following pyramid characteristics:
- Straight-sided, with generations of similar size. In the case of Sweden, we know that this reflects falling birth rates and increasing life expectancy (remember, the pyramid does not tell us this; it just reinforces it).
- Under-cut at the base. This could be because of a sudden surge in infant mortality, but we know that in the case of Sweden it is down to falling rates of fertility.
- An asymmetry between men and women in the older cohorts. This is because women have higher life expectancies than men in almost all societies.

Mexico provides a good LEDC example (Figure 5).

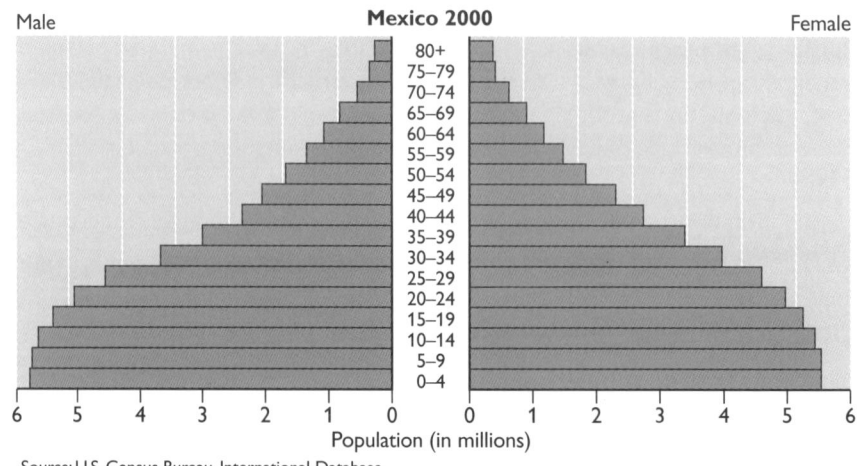

Figure 5 Mexico's population structure

This demonstrates the following pyramid characteristics:
- The wide base suggests a high birth rate.
- The relatively narrow apex suggests a relatively lower life expectancy (i.e. relative to Sweden and other MEDCs).
- A large number of young dependents (those under the age of 15) has implications for future population growth when these very large cohorts have their own children.

Variation within countries

In MEDCs, rural areas in many remote regions have lost their younger population through out-migration, giving them an older median age. (The median is the middle value in a sequence, which is better than the mean if the distribution is skewed by an extreme value. Just think about the average age of the people in your classroom when the teacher is there and contrast it with the average age when the teacher is not there. The median age gives a better description when the teacher is there.) On the other hand, some rural areas have experienced in-migration of the elderly (retirement). In some areas, this could all be confused by the in-migration of other, younger groups as well. This is the case in parts of Texas to which people have retired but which have also experienced in-migration of younger people from Mexico to work in local industries.

Urban areas in MEDCs will have equally contrasting pyramids. Army towns like Aldershot will have a structure that looks very different from Worthing, which has a very high percentage of elderly retirees. University towns and older industrial areas, such as some of the old coal-mining centres in south Wales, will show similar contrasts.

In LEDCs there is a comparable degree of variety, but the forces and factors involved are rather different. In many of the poorer LEDCs, especially in sub-Saharan Africa, the dominant theme is rural–urban migration. In some areas, this is highly selective, leaving an older and predominantly female population in the rural areas. This is also true in far wealthier LEDCs such as Mexico, which has seen out-migration to the USA, especially in northern regions.

Do not assume that rural populations are declining in LEDCs. In many cases, out-migration is more than compensated for by a higher-than-average birth rate.

The population of urban areas in LEDCs is characteristically younger, reflecting age-selective migration and the impact of this on natural increase.

Populations change over time

The components of population change, both natural (crude birth rate and crude death rate) and migration

- *Use population data, from a variety of countries in various states of development, to identify the components of population change — both natural change and as a result of migration.*

Population change — definitions
- **Population change:** the changes in population over a given period, expressed in absolute figures or proportional figures (rate of change).
- **Birth rate (BR):** the number of live births in a given time (usually 1 year), expressed as a rate (per 1000 population).
- **Death rate (DR):** the number of deaths in a given time (usually 1 year), expressed as a rate (per 1000 population). About 600 000 people die every year in the UK out of a total population of about 60 million. Thus, about 1 in 100 people die each year. Conventionally, this is expressed as 10 per 1000.
- **Fertility rate:** the average number of children born to women in their reproductive life. This figure is very illuminating. Should a woman produce two children, she has, in effect, produced replacements for herself and her partner (obviously not all women with two children have a girl and a boy, but it does even out over the whole population). If the fertility rate is greater than 2, the population is expanding; if the fertility rate is below 2, the population is contracting. The fertility rate in the UK is 1.5.
- **Natural change:** the difference between births and deaths, expressed numerically or as a rate (per 1000 population).

Factors affecting population change

There are three major factors affecting population change: **fertility**, **mortality** and **migration**. These variables account for population change over time.

Fertility is usually measured using the **crude birth rate** (CBR), which is given by the following formula:

$$\text{CBR (per year)} = \frac{\text{total number of births in a year}}{\text{total population}} \times 1000$$

At this stage, you should think about the influence of the **age structure** of a population on birth rate.

Various adaptations can bring much more accuracy to the calculation, for example by assessing the number of children born to each age group, in relation to the number of women in that age group. The total fertility can be calculated or, equally, the replacement level fertility assessed. This last measure is more widely used and is a very quick way of assessing whether or not a population is increasing or decreasing.

Mortality is often measured using the **crude death rate** (CDR), which is given by the following formula:

$$\text{CDR (per year)} = \frac{\text{total number of deaths in a year}}{\text{total population}} \times 1000$$

Just as with the CBR, the CDR can also be established using data such as infant mortality rates or data specific to life expectancy, age and sex.

Population change for any given country or region is the outcome of the balance within and between two components: natural change and net migration, where natural change = number of births − number of deaths

When births exceed deaths, the change is a natural *increase*; when deaths exceed births, the change is a natural *decrease,* although in some textbooks the term natural increase in used even when referring to negative figures.

The total change of population (on any scale other than global) obviously incorporates migration. People move very frequently in modern societies such as the UK or the USA (1 person in 7 moves every year in the US, i.e. nearly 43 million people move house every year out of a total population of about 300 million).

net migration = number of immigrants − number of emigrants
(or in-migrants)   (or out-migrants)
(or people moving in) (or people leaving)

Reasons for variations in fertility and mortality patterns and rates

- *You should understand that adjustments in social behaviour can, in turn, affect reproductive behaviour. For example, changes in the average age of marriage, for example, clearly impact on fertility rates, but you would not be expected to give the reasons for that change beyond the immediate causes of improvements in the social status of women and consequent change in their access to education.*
- *Mortality rate changes pose fewer challenges, but it is critical that you comprehend the importance of age structure in determining mortality rates.*

Mortality
Factors affecting mortality are summarised in Figure 6.

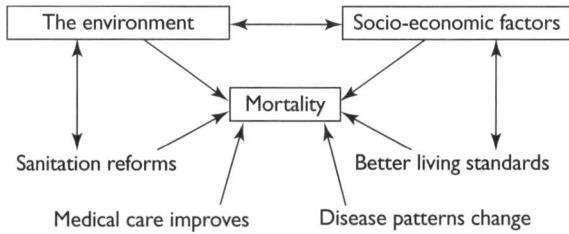

Figure 6

Fertility
Birth rates are affected by **economics**, **social factors** and **cultural factors**.

By far the most important factor in explaining variations in birth rates is whether or not people *need* children. In some societies, more children might mean more wealth. In subsistence agriculture, children from the age of about 5 often produce more with their labour than they consume. Thus, children are welcomed (especially boys — see social factors) because they can increase the production of the family.

Even in the early stages of industrial societies, when child labour was common, a bigger family meant a richer family, especially if one or more was to secure a higher-paid job.

As soon as education became compulsory for young children, when societies started to need a more skilled workforce that was literate and numerate, birth rate began to fall as children became more costly. This operates today in that poorer families will often be larger than middle-class families, while poorer ethnic groups often have higher fertility rates than better-off groups.

Wealth might also be important in allowing families to control family size should they so wish. Contraception costs money.

In many societies, the birth of a girl was no great cause for celebration. Girls would eventually move out of the family and live with their husbands, taking with them a valuable dowry (a gift) in the form of land or a few livestock. Male children, on the other hand, would marry and bring dowries into the family.

The impact of this sexist attitude to children was to increase family size as couples sought to have boys to 'compensate' for expensive girls. There are also a number of societies in which the social status of a man is still measured in terms of the number of children (especially boys) he can produce (note the importance of men, not women), and his wife (or wives) is not offered much choice in the matter. A number of surveys on contraceptive use in both South America and Africa reveal significant differences between the sexes in response to the question: 'Do you and your partner practice any form of birth control?' Unsurprisingly, not all women appear to be telling their machismo-driven partners the whole truth.

High rates of infant mortality are also very significant. If you expect some of your children to die before adulthood, you adjust your reproductive habits accordingly.

Religious beliefs are obviously important, although not perhaps as much as some would think. The Catholic and Protestant communities in Northern Ireland have contrasting fertility behaviour, with the Catholic community having a significantly higher birth rate. On the other hand, the country with the lowest fertility rate in the world is Italy — a Catholic country. Religion is obviously not a controlling factor.

The role and status of women is increasingly important. In the Kerala province of India, major reforms in land ownership and the rights of women, especially with respect to education, has led to significant falls in fertility and birth rate. In the UK and in other western countries, the increasing number of girls who go on to university has had a huge impact. A significant number then embark on a career, get themselves established and then begin a family. Another logic applies to the less well-off, who often find themselves in households that now depend on two incomes to satisfy their material ambitions. Thus, child-rearing is delayed as it inevitably takes one of the partners out of employment for a while. Thus, the average age of women when they have their first child is now 29, whereas for their mother's generation it was 24. It is not hard to see why fewer children are being born in MEDCs.

Edexcel (A) Unit 2

The demographic transition model (DTM) — its application and limitations
- *You should understand the main stages of the DTM.*
- *You need to understand that it is not an explanation of population structure.*

The demographic transition model is a simplified, visual representation of the birth and death rates in a country over the past 100 years or so. Figure 7 shows the model for the UK.

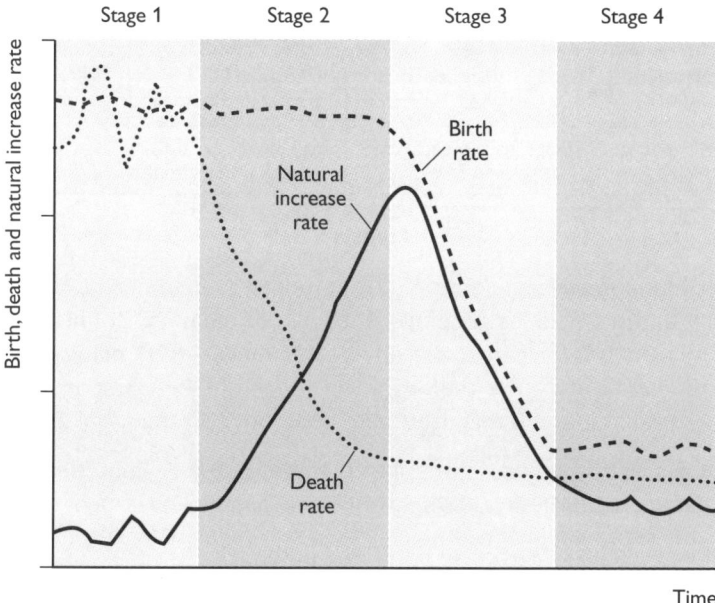

Figure 7 The demographic transition model

The stages
- In **Stage 1**, both the BR and DR are high and fluctuate. Thus, natural increase is low because the huge number of deaths cancels out the high BR.
- In **Stage 2**, the DR falls because the BR remains high. The rate of natural increase goes up.
- In **Stage 3**, the BR begins to fall, as does the natural increase.
- In **Stage 4**, the BR and DR are both low and fluctuating. Natural increase is low.

It is possible to add a Stage 5 which has been observed in a number of European countries. Here, the birth rate has fallen and, because of the increasing average age of the population, the death rate has risen.

Uses of the model
This model is a helpful way of describing the demography of the UK. A large number of MEDCs seem to have followed a similar demographic route to the UK. It is therefore a reasonable descriptive tool.

It is possible to link population structure to the demographic transition model (see Figure 8).

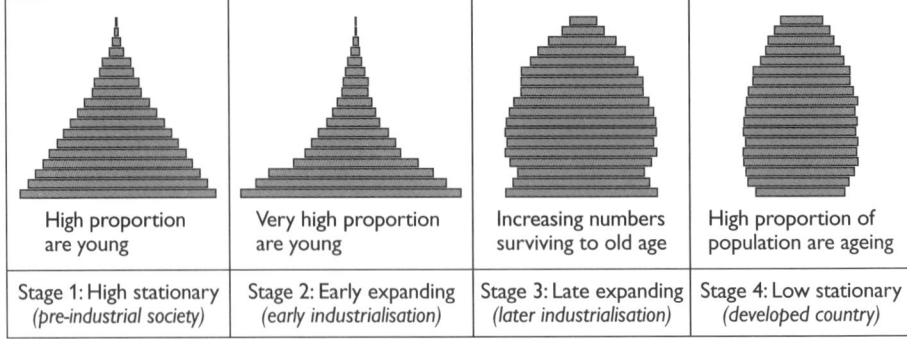

Figure 8

Limitations of the model

Even for the UK, the model is generalised. Many city areas did not fit the pattern as described by model; for instance, there were differences in BR between skilled and unskilled workers during the period of the Industrial Revolution and there are variations today in birth rate and death rate between different social groups.

The model is said to be Eurocentric, but many of Europe's countries exhibit very different characteristics from those shown on the diagram.

The model is much less descriptive for LEDCs in that:
- the DR has probably fallen in these countries for different reasons, for instance LEDCs have benefited rapidly from medical advances and packages as they arrived all in one short period
- other innovations, for example sanitation, have arrived equally rapidly, although their impact has been variable
- Stages 2 and 3 BRs in LEDCs tend to be higher because of 'young' marriages
- the populations of LEDCs tend to be larger than that of the UK, so the impact of high growth in Stage 2 appears to have been far greater
- the UK entered Stage 2 at a high level of economic growth, while LEDCs face a far greater challenge with the consequence of this stage's high growth

The demographic transition model describes what happened in the UK between 1780 and the present, but it does not *explain* those changes. The various changes in birth rates and death rates are not automatic, nor does a change in one necessarily lead to the other.

The model only looks at natural increase. It is important to remember that during the time period covered for the UK, there was a major migration overseas which, of course, had an effect on the population as a whole and on its age structure. In more general terms, therefore, don't forget the impact of migration on population change.

Finally, don't forget that all models simplify reality, though this alone is not a very illuminating criticism. You need to be able to develop it beyond a simple statement.

Changes to population structure — the characteristics of ageing and youthful populations

- *You need to be able to define dependency and be aware of the difficulty of finding a definition that can be used in countries with widely varying levels of development and economic wealth.*
- *The main theme here is the ageing of the population in many parts of the world, especially in Europe and Japan.*
- *You should be able to draw on examples at a range of scales.*
- *This strand is largely concerned with the presentation and analysis of the data, especially census materials.*

The dependency ratio

Those between 15 and 65 are known as the **active/working population**. Those under 15 and over 65 are known as the **dependent population**.

The dependency ratio can be calculated using the formula:

$$\frac{\text{children (1–14) and elderly (over 65)}}{\text{those of working age (15–65)}} \times 100$$

Generally, the dependency ratio for MEDCs is about 65–75, whereas for some LEDCs it is over 100.

This is not an altogether satisfactory measurement, given that using age as the only criterion for assessing dependency is obviously misleading. The following observations can be made:

- A large number of teenagers stay on at school post-16 and subsequently go to university.
- Parents who choose to stay at home to raise children are clearly 'dependent'.
- The unemployed are 'dependent'.
- Many retire before 65, while others often carry on well into their seventies.
- In subsistent agricultural societies, the young may work and contribute considerable amounts to production, while retirement is not a meaningful option for the elderly.

As with many other sections of the specification, you must know the definition but also recognise the weaknesses of these rather arbitrary categories.

The ageing world

With widespead falling fertility rates and, outside sub-Saharan Africa, significant increases in life expectancy, the statistics tell an obvious story. The median age is rising and it will affect your lives. You are already outnumbered by your parents' generation and given that fertility rates are at unprecedented lows in so many countries, the 'greying' of the population is bound to continue for quite a while, whatever the reproductive behaviour of your generation. The following table shows the median ages of the populations of continental areas.

	1975	2000	2025
Africa	18.5	17.5	22.0
Latin America	20.0	24.0	30.0
North America	30.0	36.0	38.0
East Asia	23.5	31.0	38.5
South Asia	20.5	24.0	31.5
Europe	30.5	38.0	41.0
World	**23.5**	**26.5**	**32.0**

The median is a better measure than the mean in distributions that are skewed because it takes the middle value in a series, not the average, which will be distorted if there are large numbers of either young or old.

Population changes have socio-economic and environmental implications

The socio-economic implications of youthful and ageing population structures, to include dependency

- *You should understand the problems and the opportunities that a youthful or ageing population structure present for governments, the economy and society on both a national and a local scale.*
- *This could include not only a study of health care, social welfare and pension provision at a national level, but also local studies of social, transport and retail provision which might be built on for AGS work in the field or as a basis for coursework.*

The 'greying' world population

The consequences of changing population structure are not all negative. You should be able to categorise them in the following way:
- social
- economic
- cultural

Be careful to establish whether an examination question asks you to examine 'changing population structure' in the context of either MEDCs or LEDCs or whether it does not specify.

In MEDCs, the old will increasingly have to look after themselves in terms of pension provision; with fewer people working, there is not the revenue to support the various state pension systems.

The 'greying effect' impacts on the young in many countries, with the maintenance of some vital sections of employment only being achieved through immigration. Fewer children means fewer babies, which means, ultimately, fewer workers. This may have serious implications for long-term economic growth. On the other hand, with more and more mechanisation apparent, perhaps a smaller but better-trained population is exactly what we need.

The influence of the older population in the UK is very considerable. Brought up to 'save', the 55–64 cohort saves 85% more than the average UK household and represents 68% of those with £10 000 or more saved in stocks or in banks and building societies. They own their homes and have a seemingly large disposable income.

On the whole, the senior citizens of the UK (and the USA) wield considerable buying power, whether it be related to holidays, films or CDs. They also hold votes in greater numbers and politicians need to be aware of their potential power when allocating resources.

The producers of films, the designers of clothes and the makers of automobiles are among the many corporate enterprises which are all aware of the purchasing power of the elderly and are gearing more and more products towards them. The 'youth' culture of the 1960s is not quite so obvious 40 years on.

There are obvious pressures on health services. The elderly are much more 'expensive' in terms of health care than the young, so more resources will be diverted into geriatric health care (looking after the elderly) rather than other types of medicine. The elderly also work both formally (legally) and informally (illegally) in many sectors, as part-time helpers for charities and as advisers. In this way they contribute to production as well as demand.

Youthful populations

Many LEDCs in earlier stages of demographic transition still support extremely youthful populations. For instance, Mexico, Ethiopia and Indonesia have structures that support 50% of the population under 15 years of age.

In places like Tanzania, the population is not expected to stabilise until about 2045, as those who are of reproductive age have to work through the system. In other words, although fertility rates and birth rates are beginning to fall, the cohorts of each successive generation are larger, reflecting birth rates in the past. Thus, although the birth *rate* might be declining, the number of children being born is still increasing.

The dependency of Tanzania's youthful population puts the country's development on hold, as the population is almost totally reliant upon agriculture, both to feed it and to drive forward development. Recent policy in this country has centred on reducing the population, in an attempt to reduce debt and pressure on the country's resources.

Countries with youthful structures clearly have to organise their distribution of resources rather differently, with a much greater emphasis on education. Sadly for Tanzania, its crippling debt burden has inhibited this effort.

Population growth in relation to resources — the views of commentators like Malthus and Boserup

- You need to understand the difference between these two theories and the limitations of their applicability.
- You need to understand how population growth has affected the environment.

Many argue that resources are sufficient in the world to support the 10–12 billion population. However, inequalities in distribution mean that many people (about

1.5 billion) are never very far away from famine and about 4000 children under the age of 1 die every day from a lack of fresh water and basic sanitation. At first sight, the question of whether there are too many people in the world seems simple enough to resolve, but in fact it is one of the most controversial of all issues. You don't need to take sides in the debate (although you can), but you do need to understand the different views.

The Malthusian viewpoint

Malthus wrote his essay on population at the very end of the eighteenth century, which was a time of great change, both economically and politically. He wanted the Poor Law abolished (this was an early form of social welfare), for he believed that by artificially prolonging the life of the poor, the 'least useful' in society would multiply (by the usual methods). This rapid (geometric) increase in population would not be matched by an equivalent growth in food supply; hence, a gap between food demand and supply would lead to famine or, perhaps, civil war. The French Revolution was the big political event and Malthus, in common with many others, was very fearful of the spread of radical ideas like democracy and liberation which were, of course, particularly popular with those who did not hold power.

Those with strictly Malthusian views today tend to be reacting on instinct to the seductive idea that poverty is a function of too many people in the world. They argue that the current shortages of food in the LEDCs are the result of overpopulation; in Malthusian terms, 'the power of the population to increase is greater than that of the Earth to sustain it'. Much less crudely, the Neo-Malthusians (Professor Ehrlich and many others) are of the opinion that the increased agricultural and industrial activity needed to feed and maintain the population will eventually lead to environmental disaster and intense financial problems.

Malthus reviewed

The initial impact of Malthus was quite considerable, but his clearly stated relationship between food production and population was undermined by the enormous surge of economic activity in the early nineteenth century. Improvements in farming and a rapid growth of trade saw agricultural output and living standards rise faster than he had predicted and, significantly, much faster than population. Hence, at the end of that century there were more people who were, by and large, better off. If there was a relationship between people and resources, it was obviously not quite as Malthus had argued. The same historical, positive correlation between population and income levels was repeated in the twentieth century, when:
- population growth was more rapid than at any other time in human history (from about 1 billion to about 6 billion)
- living standards rose globally more than in any other century

This is very powerful evidence for the anti-Malthusians.

The Boserupian view

Esther Boserup took a wholly different view of the relationship. Writing from a background in economics, she suggested that population growth has a positive impact on people in that as we approach a resource boundary (put simply, something starts to run out), we invent our way out of the problem. Thus, we 'invented' farming *because* we were starting to run out of hunted food. Population growth thus becomes something positive and, what is more, quite central to our development as a species.

Boserup reviewed
This rather reassuring and optimistic view has not gone unchallenged. The neo-Malthusians take no particular issue with Boserup in the context of explaining something of our prehistory and would, by and large, go along with the general idea. What they do dispute is the applicability of such an apparently relaxed attitude in the twenty-first century. For some, the more-people–more-wealth point, which seems to be the uncontestable conclusion of the past 200 years, ignores the environmental impact of population growth.

We are, so the neo-Malthusians believe, heading for catastrophe because we are using up world resources at a formidable rate that threatens the sustainability of the planet. In essence, it is a simple argument: the earth might technically be able to support many more people, but only by destroying the resource base. It is not just a matter of global warming; there is also desertification, deforestation, salinisation and the destruction of species.

They develop this argument by adding the warning that some environmental problems might accelerate very rapidly. For example, climatic change might take place over decades, not centuries, and it might be your children rather than some unimaginably distant descendant who has to face the problems caused by our casual waste of the planetary resources.

The conclusion of this pessimistic view is that it is the rich 20% of the world's population who consume 80% of its resources. Unlike Malthus, who wanted to control the 'unbridled lust of the poor', the neo-Malthusians identify the rich who have an 'unbridled lust for consumption' as the culprits. Most of these live in MEDC countries (especially the USA, Europe and Japan). This does not make the neo-Malthusian approach universally popular.

Concepts of overpopulation, underpopulation and optimum population

- *The concepts of overpopulation, underpopulation and optimum population refer to the balance between population, resources and development.*
- *Non-Malthusian models of population (such as that of Boserup) place little value on such terminology. At a basic level, these concepts vary depending on the level of economic development.*
- *You should recognise that this area of the specification is very controversial. It might be taught to you through examples, using an apparently straightforward example of 'overpopulation' drawn from sub-Saharan Africa, but then complicated by evaluating the role of national politics (civil war perhaps), agribusiness involvement in export crop development on irrigated land and uneven land distribution on the 'carrying capacity' of a given area.*

Overpopulation exists when the population of a country or area is in excess of that which the resources of the country can support (sometimes known as its carrying capacity), at a *given standard of living*. Although a high density of population often accompanies overpopulation, it is not the same thing. The implication here is that a reduction in the population would lead to an increase in average wealth.

Underpopulation exists when the population of a country or area is insufficient to utilise fully the resources in that area or country. The components are present to increase national wealth, but there are too few people to do so. This is not at all the same thing as a low density of population. The implication here is that an increase in the population would lead to an increase in average wealth.

Optimum population represents equilibrium: the population is such that resources are utilised fully, for the greatest economic welfare.

These definitions are not difficult to remember, but contain many traps for the unwary:
- Countries do not support themselves in the way that they once did, by their own resources. Trade and the flow of goods means that they do not have to support themselves from products grown locally or nationally.
- Thus, it is not hard to imagine that a prehistoric society might have been 'overpopulated' when there were too many people for the supply of food in their environment, but when humans invented agriculture, making the environment more productive, the idea of overpopulation would have been hard to apply.
- The concept of carrying capacity is not fixed because we (i.e. people) change the environment, for better and for worse.
- Resources can be sub-divided into several groups: natural, material (or capital) and human. The last two are clearly not fixed, in that they will not 'run out' if we use them.

Settlement patterns

The size and location of settlements vary, and distinct patterns can be identified

Physical, human and political reasons for the site and situation of settlements, and distributional patterns

- *You need to understand the difference between the site and the situation of a settlement.*
- *You need to understand the possible reasons why particular settlements were established and the reason for their subsequent growth. These reasons will rarely be the same.*
- *Your understanding of distributional patterns should include the use of nearest neighbour statistics and the description of the pattern should be both quantitative and qualitative. You need to have a command of terms like nucleated and dispersed.*
- *You should have knowledge of one region with marked physical variations, and an area of more uniform physical characteristics.*

Settlement pattern can be affected by the following factors:
- the physical environment, for example terrain and water supply controlled the early distribution of people
- historical influences, for example earlier settlement patterns, old route networks and former resource locations
- current human influences, for example patterns of employment and service and leisure needs
- levels of technology, for example mobility and transport networks
- government policies controlling settlement distribution and growth

This means that in remote areas (with poor climate and transport links) settlement is generally dispersed, with the occasional farm or hamlet developing on the odd patch of more favourable land. Where there is an impetus for growth (usually economic) and where the physical conditions favour development of a settlement, a more structured and fuller settlement pattern develops. Once developed, settlement patterns show a high level of continuity because they generate further growth in a feedback mechanism. In simple terms, bigger settlements offer more employment opportunities than smaller ones. This attracts more people, the settlement becomes bigger and so on.

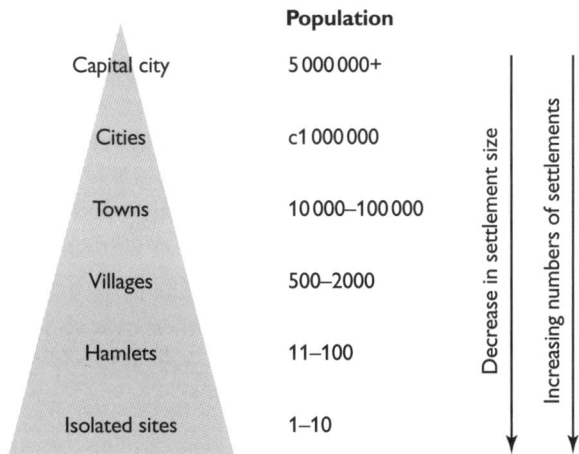

Figure 9 **Settlement hierarchy**

General features
Some areas show uniform favourability to settlement, for example East Anglia's pattern relates to a predominance of different forms of agriculture. In some areas, the quality of the soils has led to the development of arable farming with large villages distributed at regular intervals across a relatively uniform physical landscape. In some less-favoured areas of the region, the wetter clay soils have been used for dairying which leads to a more dispersed pattern of farms and hamlets rather than large nucleated villages. Cows need to be milked twice a day and so a dairy farm is surrounded by its fields.

In urban-based economies, settlement is strongly affected by urban influences; towns infringe on rural areas and there is dormitory settlement development in nearby villages. You will already have covered counterurbanisation in your study of the UK. Today, settlement pattern is strongly controlled by planning policies, for example Green Belts and New Town legislation in the UK. At a local planning level the destination of future housing development is covered by local plans, which favour some settlements (key settlements) but disallow growth in others. These planning regulations also influence the type of development.

The terms nucleated and dispersed have traditionally been used to differentiate between settlements. Highland Scotland and Wales, for example, are characterised by dispersed settlements, while lowland UK is characterised by nucleated settlements.

Site, situation and function — definitions
- **Site:** the position of a place in terms of its immediate physical location — that is, the actual land it is situated upon.
- **Situation:** the location of a place in its broader regional context — that is, the position of the settlement in relation to the surrounding land.
- **Function:** what a place does, in terms of its industrial, residential, commercial and administrative activities. Map evidence might enable us to make a few comments about a settlement's previous principal function.

Reasons for site and situation
With the exception of a few settlements founded as holiday resorts or industrial sites, most built-up areas in Britain put down their roots about 1000 years ago. Other than mining towns and villages, rarely do we find a settlement owing its location to just one geographical factor. Generally, factors combine and enable us to describe any settlement in terms of site, situation and function.

Figure 10 gives site, situation and function details for one UK settlement.

Variations in settlement size, primacy and the theoretical rank–size relationships at the national scale
- Much of this can be learnt through national population data and maps.
- The concepts should be understood at the national scale and you are advised to have a contrasting set of countries to illustrate the ideas of variations in settlement size.
- The reasons why these patterns have come about include economic, political and historical factors. Examples include federalism, recent state formation and colonialism.

Variations in settlement size
Settlements have many functions, but they can be usefully divided into two groups:
- general functions for the population of the settlement (primary schools are an obvious example). These are **city-serving** functions.
- functions for people from outside the settlement. Cambridge, for example, is widely known as a university and hi-tech city, while Great Yarmouth is a holiday resort. These are **city-forming** functions.

Edexcel (A) Unit 2

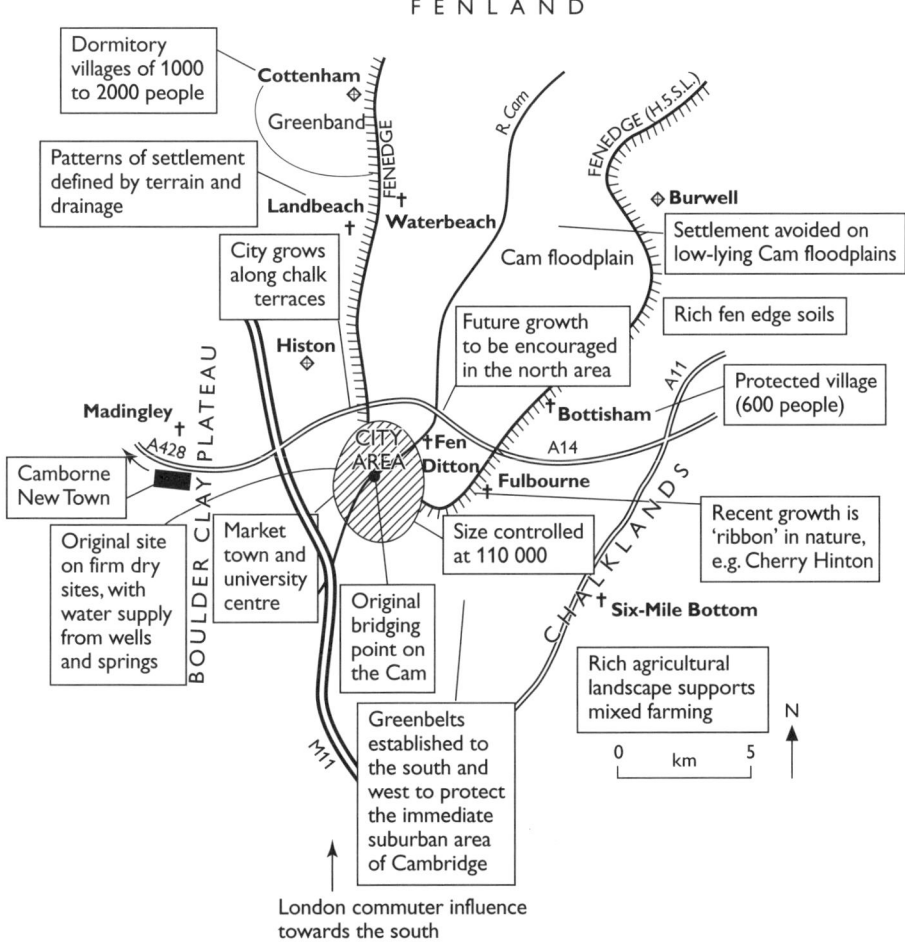

Figure 10 Factors that have affected the initial settlement and subsequent development of Cambridge

There is an obvious overlap between these groups. For instance, major shopping centres attract customers from outside the city, but also provide retail functions for the residents.

The rank–size rule

Zipf (taking on the work of Jefferson (1939): 'The Law of the primate city') recognised a pattern in the 1940s in the form an inverse relationship between the size and rank of a given settlement. This suggests that the population of the second settlement will be about half that of the largest city (the **primate city**), the third city will be roughly one third the size of the largest city, and so on. The largest primate cities are many times bigger than other settlements further down the rank and dominate the country in terms of economic and social affairs, and affairs of state.

On the whole, the rank–size rule is a descriptive rather than analytical tool. The idea of primacy is the key contribution of rank–size rule whereby, in some countries, the largest city is overwhelmingly dominant. The explanation for this varies, but often includes:
- a history of colonialism, with government and investment concentrated in the main city (e.g. Colombo in Sri Lanka)
- a history of highly centralised government, with little or no independence for the regions (e.g. Paris in France)
- a lack of industrial development, whereby a rural economy did not develop any urban centres to speak of, other than the capital city which was the administrative capital and, usually, a port (e.g. Montevideo in Uruguay)
- the positive feedback of large cities generating their own rapid growth at the expense of other cities — this is true of many of the so-called 'global cities' that have emerged in the past 30 or 40 years (e.g. Taipei in Taiwan or Seoul in Korea)
- small countries within larger trading areas (e.g. Belgium supports the large primate city of Brussels, which is a central place for many who live outside Belgium and has thus grown very fast)

Lack of primacy is often the result of relatively recent national integration. For example, both Italy and Germany became nation states late in the nineteenth century. Before that final amalgamation, each area comprised many competing states, each of which had its capital and administrative centre. Many of these developed as major centres before integration. Hence Milan, Turin, Genoa, Naples and Rome were all 'capitals' and primate within their own states before the formation of the modern Italian state. This is often reinforced by the subsequent system of government allowing much more independence to the individual regions of a country. This federal system is present in the USA, Australia and Brazil as well as Germany and Italy. This promotes the development of larger-than-average regional capitals.

Large size can also reduce the chance of primacy. The USA is a continental-sized country. It is hardly surprising that no one city dominates the urban system. The same is true for Brazil.

Settlement hierarchies — central place, range and threshold
- *You should know Christaller's assumptions, which are a useful starting point for introducing central place. For example, if East Anglia is used, you should be aware of the presence of important central places within the area (e.g. Cambridge and Norwich) or, indeed, outside the area (i.e. London).*
- *The concept of threshold population should also be understood.*
- *You must be able to describe the hierarchy of settlements and services in a given area.*

Christaller's central place theory (1933)
Central place theory (CPT) depends upon the concept of functional interdependence/importance. Figure 11 illustrates a UK example.

Larger settlements often have more than one city-forming function, but one common to all large settlements is their **central place** function. They act as market places for their surrounding regions or areas.

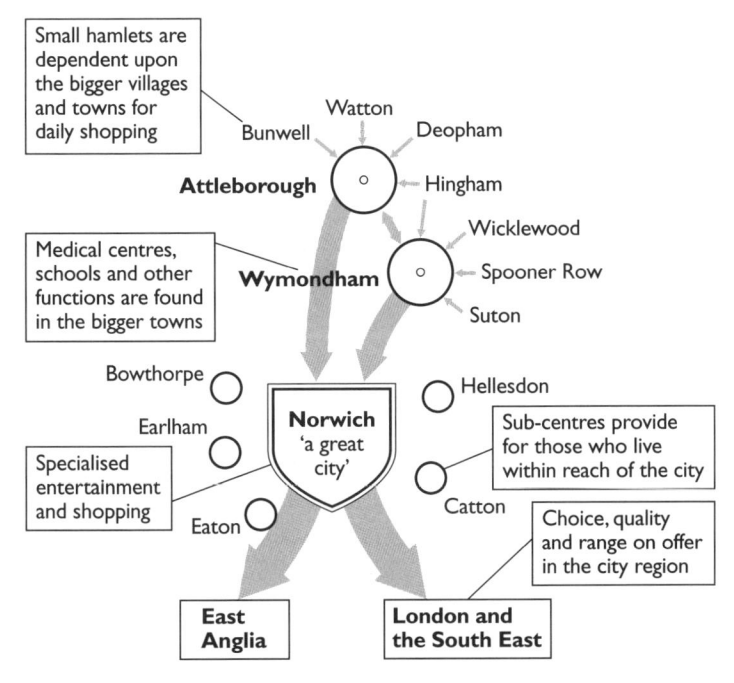

Figure 11 Functional interdependence around Norwich

Terminology
- **Central places** are settlements that develop to provide a surrounding region with specific market function. Their relative importance is measured in terms of the numbers and varieties of goods and services provided for the people who live there.
- The **threshold population** is the minimum number of people needed to support an outlet in a central place.
- The **range of a good (or service)** is the distance of the boundary (the sphere of influence) needed to enclose the threshold population for the outlet that provides it, i.e. the distance measured in time and cost that a customer is prepared to travel to obtain a good or service. For a cheap and often-required item like a newspaper, not much time or money will be expended travelling to obtain it. The range of this **low-order good** is therefore small. This means most central places, villages, towns and cities will have shops that sell such goods. **High-order goods**, by contrast, cost more, are required infrequently and people travel some distance for them. Catchments and therefore spheres of influence for settlements providing high-order goods tend to be large. Only the largest central places carry high-order goods.

The result of all of this is that a hierarchy of settlements is established if the underlying principles of threshold and minimum range are adhered to. Christaller adapted the spheres of influence that form, to hexagons, so that no gaps are left; each central place has to be accessible from all places.

The key idea about hierarchies is that they are stepped. For example, Christaller showed that each market town in a region would offer the same number of goods and services and that if there were 48 market towns there would be a predictable number of larger regional centres, 16 in the most commonly used system. These regional centres would offer all the goods and services available in the market towns but also a range of higher-order goods and services. Once again, each regional centre would be much like any other regional centre.

There is an apparent contradiction here with rank–size rule which, you will remember, ranks the populations of towns and cities in a country and rarely shows a hierarchy that would be represented by having 48 market towns all of roughly the same populations (say 20 000) and 16 regional centres all of roughly the same, much larger populations (say 50 000). In other words, intermediate settlements should not exist. In fact, however, there is no contradiction, as Christaller recognised that most towns and cities have more than one function. Hence a town might be both an industrial city, a tourist centre and a central place, and therefore it would be larger in population than an equivalent central place without those other functions.

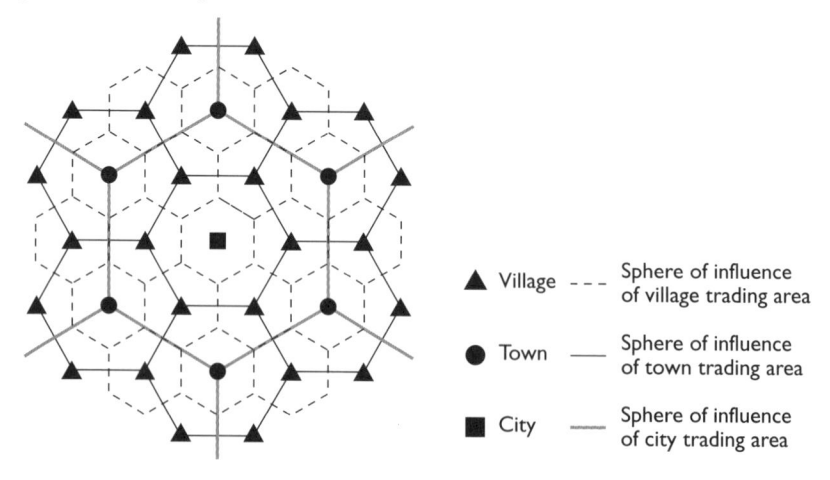

Figure 12

Advantages and disadvantages of CPT

Christaller's CPT provides a framework for explaining the relationship between centres and confirms that there is stepped hierarchy of retail centres. It does less well in describing the pattern of central places on a map because no surfaces conform to the 'isotropic' plain that was an initial assumption of Christaller's work. Where such plains do occur (e.g. southern Germany, East Anglia and the Great Plains of the USA) it works well, generating a pattern of market towns at regular intervals, determined by the distance that could be travelled (there and back) in a day using a horse and cart. But transport systems have changed and so has retail behaviour.

Out-of-town shopping centres have transformed the geography of retailing, as did supermarkets in the 1960s. People do travel miles to buy a newspaper because it is at the bottom of a shopping trolley heaving with other 'low-order' goods that are now bought in bulk. At the same time, antique shops and gift shops selling high- or middle-order goods open in small villages; these are often run by 'retired' people or those with another income coming into the house. Thus they do not 'profit-maximise' and seek the best possible locations for their enterprise. It is not possible for central place theory to cope with these common deviations from 'logical' economic behaviour.

Theoretical and practical ways of determining spheres of influence

- *You need to be able to determine the sphere of influence of a town, using secondary sources and primary research.*
- *Useful comparisons could be made with the notional predictions of gravity model-type calculations. You should be aware of the weaknesses of such models.*

Christaller demonstrated the influence of settlements for a distance over their surrounding area. His theory gave way to attempts at *calculating* spheres of influence or urban fields, with **gravity models** enabling such calculations to be made.

Reilly's law of retail gravitation

Between two towns there exists a breaking point — people living on one side shop in one centre (call it town A) and people on the other side shop in the other town (call it town B). Reilly devised a formula, using population size, to demonstrate the position of this breaking point:

$$\text{breaking point distance from town A} = \frac{\text{distance between towns A and B}}{1 + \sqrt{\frac{\text{population A}}{\text{population B}}}}$$

By calculating the breaking points of several towns of the same order around a larger settlement, the sphere of influence can be drawn in.

Problems with Reilly's model include the following:
- People don't always act in a logical way.
- It assumes too much.
- The advent of the car has changed people's shopping habits completely.

Huff's model

This was an attempt to predict movement using probability (*P*). Huff's assumptions were that shoppers:
- compare the ease with which they can reach different towns
- look at the number of shops in rival towns and compare them with their 'home' town

The formula is:

$$P = \frac{\text{number of shops in centre A}}{\text{total number of shops in whole study area}} \div \frac{\text{distance or travel time to reach them}}{\text{total distance or travel time to reach them}}$$

Problems with Huff's model include the following:
- It does not consider the wide range of factors that influence decisions.
- In reality, people do not make rational decisions.
- There are instant flaws in its usefulness at predicting since the quality and availability of goods and services are not necessarily uniform from centre to centre.

Usefulness of the two models
Both models provide a good starting point for an examination of retail behaviour.

Nearest neighbour analysis
It is often difficult to describe the patterns of settlement that are seen in the landscape. Nearest neighbour is the objective measurement of 'pattern'. However, it only looks at point patterns when three key distributions are in fact recognised: point patterns (clusters), lines (regular) and areas (random). Nearest neighbour does not explain observed patterns, but compares observed patterns with assumed or predicted patterns. Values of nearest neighbour range between 0 (clustered) and 2.15 (regular).

Problems with nearest neighbour analysis include the following:
- The area chosen to be investigated can affect outcome.
- It only really measures relationships between one individual and its nearest neighbour.
- It does not consider the direction of the nearest neighbour.

Settlements vary in their internal structure

Spatial variations in land use patterns in urban settlements — retail, commercial, industrial, recreational and residential
- *Using your chosen urban area, you should be able to identify various types of land use and be able to plot them on a map.*
- *Different types of residential land use might be identified using ward level statistics.*
- *You should be aware of different methods of grouping the data using, for example, types of ownership on the one hand and socio-economic data on the other hand.*
- *Comparative studies of MEDC and LEDC urban areas should concentrate on descriptive differences in the distribution and extent of varying land use categories. They should include descriptions of functions (residential, industrial, retail, commercial and recreational) and the nature of the resident population, including social, economic and ethnic characteristics.*

The development of cities in MEDCs
Between 1875 and 1920, cities grew dramatically in land area and in population. Land uses within the city became differentiated by area. Increasing agricultural surpluses released more farmers from their fields. The demand for labour in the city was high. In-migration and rural–urban migration was very high and most MEDCs were becoming urban nations.

Changing technology influenced city form as well as city location and city growth. Most important among these new technologies were trams, electricity, assembly lines, gas and electric lighting, street lighting and residential lighting, steel frame construction and the elevator (the latter two making possible the skyscraper).

Driving the new urbanisation was vastly expanding industrial production. Many countries were moving from traditional agrarian societies to modern market economies.

Mass markets were generated by the separation of production and consumption, which used to take place under the same roof. New technologies were generating new products as well as new ways to manufacture them. New products and expanding markets were leading to greater employment and therefore to higher incomes, more consumption and more profits.

Differentiation within the workforce was becoming more defined. Production lines and large factories led to different skill and managerial levels and differing incomes for workers. Part of the higher incomes and new wealth was taxed away to build infrastructure in order to support the cities. This included roads, bridges, tunnels, sewers, water lines, schools and street lighting.

There was still a great deal of conflict between various users of land in the city due to the lack of zoning or other land use controls. Private property rights still ruled in 1875, but by the end of the period had begun to give way to community pressure and zoning.

This period set the template for the modern city — much of the land use and the built environment that we see today stems from that period.

One of the most important changes which occurred during the period was the specialisation of land use by type. Remember that in the early nineteenth century city, all land uses were mixed together. While these had started to separate out in the period between 1840 and 1875, most of this spatial separation occurred after the introduction of railways and the omnibus as the means of moving people in and through the city.

The importance of the rail system in shaping the urban core was vital. After the railways, the major competitors for space were industries, governments, warehouses, retail, wholesale and residential. Fortunately, these users did not all want exactly the same space, but they all preferred centrality. This was the basis for what became bid rent theory. Since there was no planning to control where particular uses went in the city, those who were willing to pay the most, got the best locations. Retailers were willing to pay the most for the more central locations because they needed to be within walking distance of the rail terminal in order to be accessible to all. Their revenues and, therefore, their profits depended on the number of people who could walk in from the station.

Manufacturers were able to pay for centrality but did not value it as highly as retailers because they did not have the same intensity of use of space or, put simply, they could not make as much money per square metre of space. They were more likely to want a site on the water, if such existed, or along the railway lines — but not too far from the terminal because some of their labour force, especially managers, had to come from the terminal.

Warehouses were also dependent on the railroad for shipping but, unless their product was perishable or very high value per mass unit, they did not need a great degree of centrality. They could not bid as much for land in the CBD as manufacturers or retailers, so they located on the fringe of the CBD, along rail corridors.

Economic, political and physical reasons for variations in land use — accessibility, bid rent curves and peak land value

- *You must establish the key physical factors that might constrain urban development and then overlay this with social and economic patterns.*
- *It is important to appreciate that urban geography is complex and dynamic and that all attempts to categorise are, at best, limited.*
- *Bid rent theory should, primarily, be used as a way of explaining a process which could lead to a better appreciation of the real patterns generated in urban areas.*
- *The role of local and structure plans would be central here and act as useful counterweights to the free market assumptions of bid rent theory.*

Bid rent theory and other influences on land use patterns

This is an attempt to assess theoretically the role of locational rent in shaping the urban land use pattern. It is based on the following assumptions:

- Land in itself has no value. The price that people are prepared to pay for it is controlled by the use to which they can put it and, more particularly, how much revenue they can generate from it. Thus, selling sweets generates a given revenue per square metre per year, while selling computers generates a different income.
- Some types of land use are more intensive than others. One small office of only a few square metres might generate considerable income from the six or seven people who work in it exchanging foreign currencies or moving money about the world. Similarly, a busy high street store might generate a high income from a relatively small space.
- Different land users have different requirements. Office and retail activities may require central locations where they are accessible to customers and to workers. Companies might be prepared to offer high rents given that they would expect these locations to generate higher incomes. Further from the centre, they would not expect the same level of revenue, so to maintain their profit they would reduce the level of rent that they are prepared to pay.

The **bid rent curve** thus shows the amount of rent that different types of land user are prepared to pay for potential sites. Broadly speaking, retail and office (known collectively as commerce) will pay highly for central sites, whereas other users are more indifferent to the attractions of being central and so will bid relatively less.

There are several problems in the application of bid rent theory to explain patterns of urban land use. These include:

- the absence of a free market in land (in most countries, planning does not allow this).

- the lack of distinction between low-order and high-order services. Bid rent theory suggests that all retail activity will locate centrally while low-order functions, such as newsagents, clearly serve local markets.
- changes in the demand for space. The willingness to pay rent will vary over time, but location is subject to inertia. You cannot just change location; it is not without costs.

Models of urban growth and structure — their application and limitations

- *Each city has its own unique arrangement of land use. In MEDCs, the city is dominated by retail activity and the suburbs, probably, by residential use.*
- *The structure of a city is a combination of form and function. As any settlement develops, functional zones start to appear (these are easily recognisable on maps). The form and arrangement of these functional zones are together referred to as a city's morphology.*
- *Models have been developed to attempt to find and describe patterns of form and function. It is worth remembering that all models are simplifications of reality and that all city areas are dynamic and unique.*

Definitions

- **Central business district (CBD):** the social, commercial and cultural focus of a city.
- **Zone of transition:** an area of mixed, poor-quality, older housing, car parks, slums and light industry, which surrounds the CBD.
- **Residential housing:** any lived-in property from terraced housing to high-class residences.
- **The commuter zone:** beyond the built-up area, mostly made up of dormitory housing, golf courses, public utilities etc.

Burgess's concentric model

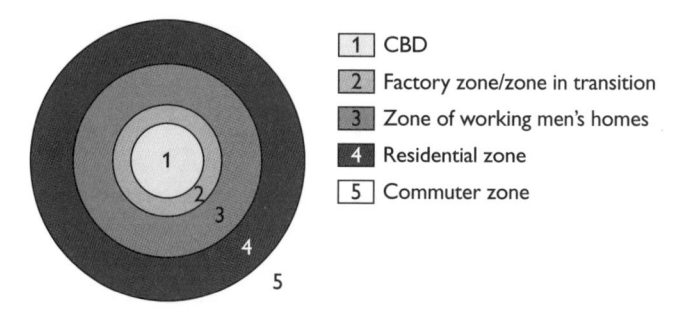

1. CBD
2. Factory zone/zone in transition
3. Zone of working men's homes
4. Residential zone
5. Commuter zone

Figure 13 Burgess's concentric model

According to Burgess:
- the city grows because of immigration to the centre; thus the model presupposes that the city is growing from the inside outwards
- social class and income levels increase with distance from the CBD, as older generations are displaced outwards by the new arrivals

- the CBD is dominated by commercial activity
- population density peaks in the high-density, low-cost housing zone close to the centre and declines outwards

Burgess's model raises the following criticisms:
- Such clear-cut boundaries are hard to find in reality.
- Only ground-floor use is considered.
- The model is based on Chicago in the early 1920s and not all cities have experienced this history of rapid in-migration.

The Burgess model is an ecological model. It suggests a particular geography of cities, but is more concerned with the processes involved. It is an **ecological** model because:
- the new arrivals *invade* the inner-city areas
- they displace the previous immigrants through a process of *succession*

Hoyt's sector model

1. CBD
2. Wholesale light manufacturing
3. Low-class residential
4. Medium-class residential
5. High-class residential

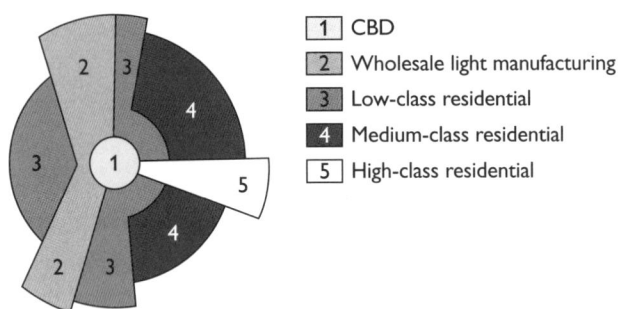

Figure 14 Hoyt's sector model

According to Hoyt:
- the model emphasises the role of transport and its relation to sector development
- certain activities deter others through the forces of attraction and repulsion
- better housing is away from industry

Hoyt's model raises the following criticisms:
- It is based on a rent study rather than a land use study.
- There can be variations in sectoral land use.

The two models are not in conflict. The Burgess model traces the development of cities during a period of rapid in-migration. Hoyt suggests that some of this expansion follows routeways and, based on his measurements of rents in US cities, identifies a tendency of given sectors to expand outwards, reflecting forces of:
- **repulsion**, whereby negative features such as railways or industrial areas are avoided by higher-class residents
- **attraction**, whereby positive features such as riverside locations or high ground with views become the focus of outward development of higher-class housing

It is not surprising that most UK cities do not fit these models very comfortably, given that they have not experienced the same processes of rapid twentieth-century growth by in-migration of distinctive ethnic groups, as American cities did. The models should be treated as a way of examining processes rather than worrying too much about whether your local town conforms to one or other model.

Urban development in LEDCs
Remember that most LEDCs have grown more rapidly in recent years than MEDCs. Cities grow in a different way in LEDCs from those in MEDCs. The most obvious difference is that new arrivals have not had jobs to go to in the inner-city area, but have arrived on the margin of the city more in hope than in expectation of something better than the rural misery from which they have escaped. Like hailstones, LEDC cities have mostly grown by accretion, with layers added to the outside.

Identification of urban functional zones
You are expected to be able to identify various types of land use and to be able to plot them on a map. This identification could be through the use of census ward data (such as housing type information, ethnic groupings and socio-economic data) or through the use of data gathered through fieldwork. You might also be expected to identify such zones from a map in the examination.

How to identify land use zones on examination maps
- Identify the main roads, railways and other important communication routes.
- Find the town/city centre. Clues include the location of cathedrals, route foci and town halls.
- Mark on the town centre or CBD.
- Identify industrial areas. Look for the word 'works', for instance.
- Identify the terraced housing. Gridiron networks and straight rows of back-to-back housing are what you are looking for.
- Open spaces, golf courses and major leisure areas can be identified.
- Look for newer areas in the suburbs. They have sweeping road networks and cul-de-sacs.

It is at this point that you usually have to explain the pattern you have discovered using the evidence that you have identified. This is a technique you need to practise on a map of your own home town or city.

Physical influences
Industry is often sited near rivers, sometimes on land that is liable to flooding and is thus not suitable for residential use. Land off the floodplain, which is better drained, is associated with higher-quality housing.

Lower-quality housing is found on lower land, while steep slopes are, on the whole, left as open space, although this very much depends on the level of demand for land. Thus, in Hong Kong steep slopes are used intensively.

Higher ground is often preferred by residents who enjoy the view or the relatively better air quality.

Coasts and lakesides also have a clear impact, as zones of land use are bisected. The

CBD is usually at or very close to the coast, with tourist facilities stretched along the coast, away from the centre. Hotels follow this zone, with the residential areas further away from the centre in broad zones that follow the shape of the coast. Many of the UK's resort towns duplicate this 'model'.

Social factors

In many urban areas, the age of the housing defines the social status of the inhabitants, with lower classes to the centre, and middle and higher classes towards the urban margin. The construction of social housing in large estates towards the urban margin distorts the ideal pattern of residential zones by introducing lower-class housing into areas of higher- and middle-class housing.

Land ownership can be important, as the existence of large open spaces in urban areas is often a result of landowners preventing development of the land. These land owners include the Church of England, private landlords and local councils.

Planning, which has been applied in the UK since 1945, tends to reinforce the existing land use and prevent major changes. Recent legislation, however, has backed wholesale redevelopment projects under the auspices of the urban development corporations, as well as local and regional structure plans.

Settlements change over time

The nature of changes in urban areas, to include function, land use, street patterns, building age and height, population charcteristics

- *You will already appreciate that urban areas are dynamic; this dynamism should be explored in some detail in this strand.*
- *The change in dominant function could be the transformation of a fishing town to a tourist area (e.g. St Ives in Cornwall). Land use changes within an urban area could include the refurbishment of industrial areas (e.g. Trafford Park in Manchester) or spatial change in land use from the city centre to the periphery.*
- *You should be able to identify street patterns from OS maps to identify various ages of settlement.*
- *Building age and height should be described and explained using a local or LEDC example. For example, in Hong Kong most buildings are multi-storey even in the periphery because the bid rent of the land is very high. Some low-rise houses are found on the Peak and in the New Territories.*
- *You should have a sense of place throughout your descriptions — the more detail you can provide the better.*

Settlement patterns are never static for long. MEDCs have experienced change over many years. This change has been dominated by movements of population out of the city (counterurbanisation) and back into the city (re-urbanisation). Redevelopment, especially in and around the city centre, has been the response of officials worried by the apparent loss of the populace. The response of planners to the above processes has focused on changes both outside and inside the city. This section deals with the latter.

Definitions
- **Suburbanisation** is the decentralisation of people, employment and services from the inner part of the city towards the margins of the built-up area. The effects of suburbanisation are felt within the city and in the rural areas around major cities.
- **Counterurbanisation** is a change that really extends beyond the city area. In terms of population concentrations, the process of urbanisation appears to be reversing in some MEDCs. In other words, people are leaving the city and moving to smaller towns and villages because of improvements in transport and ICT, government policy, perceived differences in quality of life and office relocations. This process affects both the shape and form of settlements.
- **Re-urbanisation** is the return of the people to the city. This has an obvious effect within the city.

Reasons for changes in urban areas, including zones of transition and suburbanisation
- *You should know the reasons for changes in urban areas, including the processes operating in zones of transition and residential areas, in both LEDCs and MEDCs.*
- *The reasons should be organised into various categories, such as economic, social and political.*

Redevelopment and change in the city
In most MEDC cities in the 1990s, it was recognised that there was an urgent need to revive and redevelop flagging city/central city areas. This was a response to the changing world economy; globalisation was having the effect of switching our employment structure, with a shift from manufacturing industry to service industry. Most plans involved aiming to stop the loss of population and employment, improving the housing stock (including removing the worst residences) and generally upgrading the city image. But, more importantly, plans had to attract the mobile investment created by globalisation.

Many cities have responded to the challenge. For example, Birmingham's redevelopment of the city centre was aimed at the creation of a safe, profitable and pleasurable environment. Through a so-called growth coalition, several flagship schemes were promoted. The principal area of development was in the derelict areas in the northeast part of the city — the Heartlands initiative aimed to develop the office space within the city and to return some housing. Other city projects involved building an international convention centre, the National Indoor Arena and various developments in and around the Broad Street area.

The changing CBD
The cost of a central location and shortage of land mean that there has been an increase in the height of the buildings to cope with demand. There has also been a slow, but remarkable, change in the format of the CBD. Increasingly, it is going underground and into shopping malls. At street level, pedestrians are being separated from vehicle movements. Furthermore, many operations have left the CBD to the

banks and large department stores that can afford the location, opting for more convenient and profitable sites away from the city centre (e.g. Norwich and Bristol).

The transitional zone
For many decades, the transitional zone was characterised by a mix of housing and industry. Today, however, these areas are extremely diverse. There is pressure on this zone, particularly where it is in close proximity to higher-status areas of the CBD. On the whole, industry is being driven out, and overflow commercial activity from the CBD is moving in (so-called active assimilation); industry that is left tends to be the 'footloose' IT and high-tech firms. Where the zone intersects with lower-status activity, it tends to retain much of its dilapidated housing and traditional small industrial units and warehouses. These areas are passively being assimilated into the CBD (e.g. London and Nottingham).

Commercial activity in the suburbs and on the city edge
The hierarchically ordered sub-centres within the city (that is, the corner shop, shopping parades and small clusters of CBD-type activity) are changing. As cities have grown, so have the numbers of district sub-centres serving specific areas of cities. These areas attract hypermarkets and supermarkets, their emergence triggering the growth of residential areas. Corner shops in many areas are now a thing of the past, being outgunned by both district supermarkets and the forecourt shops attached to garages.

Out-of-town shopping and the relocation of offices
This has been the principal feature of most recent urban development. The growth in service industry has caused unprecedented demand in the city; the costs of these central city sites have rocketed and, naturally, entrepreneurs have looked elsewhere. It is to the margins of the city that companies have moved. For example, Gravesend in Kent has changed markedly, with the abandonment of the High Street and with the development of the Bluewater retail park.

Manufacturing in the city
Growth causes cities to expand their industrial base (both manufacturing and tertiary employment). This then gives rise to an extremely complex locational pattern. Originally located close to the city's core and near routeways, industry is increasingly having to move to areas of the city where land is cheaper and where road transport can easily reach it. This usually involves a peripheral site, such as a business park. Industry has effectively suburbanised too!

Reasons for, and issues associated with, edge-of-town development and city centre redevelopment
- *You should know of the differences between greenfield and brownfield development and be able to evaluate the advantages and disadvantages of both types of development.*
- *Out-of-town developments need to be studied through an example. You should include the reasons for the project and the effect on the local environment, as well as wider economic, social and environmental impacts.*
- *City centre redevelopment should be examined in terms of both the motivations and the consequences of the project.*

Swindon and Nottingham — two examples of changing urban areas

The home of railway engineering for many years, Swindon now has a diverse range of other industries based on a mixture of brownfield and greenfield sites. For instance, the old railway yards have been redeveloped by the Swindon Development Agency into a range of small engineering businesses and outlet sites. In developing these, good access and high standards of building have ensured all sites are fully leased. With its excellent 'M4' position, local government support, an educated and skilled workforce, a good infrastructure and pleasant residential surroundings, Swindon has been able to attract a number of the bigger players onto greenfield sites around the city. The biggest and most influential is Honda UK, which now employs several thousand Wiltshire people in a 365-acre state-of-the-art car plant, near South Marston.

Like Swindon, Nottingham has experienced rapid expansion of out-of-town shopping. Although the effect on the city centre has not been marked, it is likely that future development will be severely curtailed by local legislation. Traditional activity in the core area of the city has been replaced by commercial office space, entertainment facilities and some specialist retailing operations. The land uses have shown an increasing tendency to cluster. Like Swindon, the dominant move has been towards the proliferation of service activity. On the whole, activity is still confined to the River Trent Valley.

Characteristics of inner cities

In the UK, inner-city areas tend to comprise older residential properties and industrial activity lying between the CBD and outer areas. They are frequently referred to as transitional zones. Inner-city areas have experienced:

- **population decline**, because of improvements in transport, slum clearance and redevelopment, the attraction of urban lifestyles and serious job losses
- **economic decline** — there have been massive factory closure programmes in inner-city areas since about 1951 and the service revolution has done little to halt the drift from the city. The reasons for this move from the city include lack of space to expand, unattractive surroundings, outdated industry, roads not being up to the demands of today's industry and obsolete buildings. Where industry has returned — whether service or manufacturing — women have taken the bulk of the jobs on offer.
- **physical and environmental change** — the inner city contains a mix of both old and new buildings. Nineteenth-century buildings tend to be in a chronic state of disrepair and, where used, are often privately owned and overcrowded. Newer housing is generally confined to tower blocks, with their range of inherent problems.
- **increased levels of poverty and deprivation** — inner cities exhibit all the socio-economic factors one might associate with areas in decline. They contain more unskilled workers than is normal, large communities of New Commonwealth residents, many single-parent families and a large proportion of long-term unemployed.

All the above factors can be seen in the inner cities of the largest cities of the West, including those in the UK. In Birmingham, for example, the city areas of Aston, Ladywood, Sparkbrook, Smallheath, Washwood Heath and Stockton Green have a stranglehold on the city centre/Nechells area.

Explanations for inner-city problems
- **The quality of the built environment** (1940s to 1950s) — some suggest that the run-down nature of the inner city was its principal problem.
- **Social deprivation** (1960s to 1970s) — this relates to the so-called **cycle of deprivation** (i.e. low wages–unemployment–overcrowded housing–ill health–poor school attendance–poor skills and so on). Social culture means that antisocial attitudes are passed on within families.
- **Planning failures** — there have been many building and planning failures in the inner city. Many sites are still derelict, while others are just poorly built. Examples include the sink estates of high-rise housing around the UK.
- **Economic and political change** (1970s onwards) — the restructuring, reorganisation and rationalisation of industry, nationally and internationally, has had a massive impact on the inner city.

Programmes aimed at improving the inner city include the establishment of enterprise zones, urban development corporations and projects like the Action for Cities Scheme.

Impact of physical site on settlement growth and expansion
- *You should know how physical factors have played a part in controlling the nature and rate of development of urban areas.*
- *Thus, the particularities of an individual site will not only determine the original nature of an urban area but also control its subsequent growth.*

You should be able to apply these points with reference to your chosen case studies. It is obvious that the constraints of physical site will be significant in the development of almost all urban settlements. Burgess's model was of Chicago, a lakeside settlement, but you should not assume that site factors are less important in your local examples. The impact of the following could apply:
- altitude
- rivers/floodplains
- coasts
- relief/slope angle

Population movements

Population movements can be classified
Population movements, to include a range from short-term circulations through to permanent migration, and from voluntary to forced
- *You should be able to distinguish between forced/voluntary, internal/international, temporary/permanent or various combinations of rural/urban movement.*
- *You need to be aware of some of the problems involved in categorising migration. Thus, Irish migration to the USA in the 1850s or Kosovan migration to Macedonia in the 1990s was 'voluntary', although the constraints on choice were considerable.*

- *In the same way, the definition of permanent can only be judged by outcome, in that many 'permanent' migrants return to their areas of origin, despite their original intentions.*
- *Many migrants have temporary status (guest workers are a classic example) but are, in reality, permanent. Others might have the intent and the right to stay, but ultimately return (one-third of all African rural–urban migrants return to their villages within a year).*

The first distinction to understand is that between migration and circulation. **Circulation** takes place when there is movement on a daily, weekly or even seasonal basis, but the permanent place of residence does not change, for example shopping trips, holidays or commuting. **Migration** involves a change in the place of residence.

Because different types of migration have quite different consequences, special terms are necessary to distinguish between different types of migration. The first set of terms distinguishes between movement *within* countries and movement *between* countries.

Internal migration

This refers to human movements *within* a nation. It usually results from people moving from:
- rural areas to live in urban areas (rural–urban)
- smaller urban areas to cities (urban–urban)
- between cities (urban–urban)
- cities to rural areas, where new industries create new jobs (urban–rural)

A good deal of internal migration is **permanent** as families move and never return to their birthplace. This is the hardest to define, given that people may have the intention of staying for the rest of their life but cannot know, for sure, that they will.

Some forms of movement are hard to classify. Important economically, but not demographically, is the very common movement that involves people in MEDCs moving house because, for example, their family has expanded. In this sort of move, there are few changes in terms of such factors as jobs, schools or social life. In the USA and the UK, about 1 in 8 of the population moves every year, but well over 50% of these movements are local and have no profound impact, except on the housing market.

All of this internal migration can be very costly in terms of demands on services (e.g. schools in one area might need to be expanded) and the infrastructure (e.g. roads in one area becoming busier).

International migration

This refers to human movements *across* national boundaries. This form of migration has become more and more common since the middle of the nineteenth century, as more and more people have moved in search of new opportunities and safer places in which to live.

Much international migration is known as **economic** or **labour migration**, involving **migrant workers**, and occurs as people move from countries with lower wage levels to countries with higher wage levels in search of work. Some international migration is known as **family reunification** and happens as members of a family move to join other members of their families who have already settled elsewhere. **Refugee movement** occurs as people move to escape wars and political or religious persecution.

Governments do try to regulate international migration. The governments of **receiving countries** attempt to regulate immigration flows, to ensure that most of those people who enter a country:
- have skills that can be used in the economy
- will find work
- will contribute to the country through taxation

The governments of **sending countries** may also try to regulate emigration flows to ensure that:
- valuable professional and technical skills, sometimes known as human resources, are not lost to the country's economy
- emigration does not result in falls in agricultural production and a loss of food security
- labour shortages in the country do not prevent new forms of economic activity and growth

The regulation of out-migration by sending countries is much more difficult because migrants usually think of their own personal interests and their families' interests when they make decisions about moving. They do not usually think about the consequences of their movement for the national economy or for national development.

Application and limitations of gravity models in predicting migration flows
- *You should be able to carry out gravity model calculations.*
- *You should be aware of the uses and limitations of this model and understand other models and theories, such as those of Ravenstein, Lee, Huff and Stouffer, which could be used to illustrate basic concepts of human movement.*

In the 1880s, Ravenstein put forward what have been termed the laws of migration, based on observation of patterns in Britain, and supplemented by data from the USA:
- Most migrants travel short distances. This is still true today — most movement in MEDCs is local.
- Migration proceeds step by step in that the majority of movement involves stops along the way.
- Longer distance migrants prefer to go to great centres of commerce or industry.
- Each stream of migration produces a counter-stream of people moving the other way.

- Urban dwellers are less likely to migrate than people in rural areas.
- Females are more likely to migrate than males in internal migration, but movements of males are more common in international migration.
- Large towns owe more of their growth to migration than to natural increase.
- The volume of migration increases as transport improves and with the development of industry and commerce.
- Most migration is from the agricultural areas to centres of commerce and industry.

These laws emphasise the importance of the characteristics of origin and destination. The gravity model allows us to measure whether volumes of migration are inversely related to distance, as Ravenstein suggests — a phenomenon known as distance decay. The simplest form of gravity model can be expressed in the formula:

$$M_{ij} = \frac{P_i P_j}{D}$$

where
- M_{ij} = migration between two places, i and j
- P_i and P_j = population sizes of i and j
- D = the distance from i to j

It assumes:
- each migrant has the same information
- movement costs are the same in all directions

It is evident from this model that migration will take place more frequently between larger places than between smaller places and that the greater the movement between the two places, the less movement there will be (two more of Ravenstein's Laws!).

The gravity model can be modified by the way we measure 'size' (this could be numbers of workers or particular types of mover, e.g. retired) and 'distance' (straight line, by road or in terms of time or cost).

Stouffer suggested some improvements to the gravity model. He put forward the concept of 'intervening opportunities', i.e. that a number of alternative destinations exist between and within the origin and destination. So, for example, a person at i moving to j could consider other destinations.

Stouffer suggested that the level of movement between two places is dependent on the number of **intervening opportunities** between them. Intervening opportunities are the number and characteristics of other migration destinations which may exist between place A (migration origin) and place B (migration destination). The main feature of this model is that the nature of places, rather than distance, is more important in determining where migrants go. People will move from place A to place B based on the real, or the perceived, opportunity at place B (e.g. work). According to Stouffer, the number of people moving over a given distance is directly proportional to the number of opportunities at that distance, and inversely proportional to the number of intervening opportunities.

Lee tried to explain the factors affecting migration in terms of the **positive** and **negative** characteristics of both the origin and destination. Migrants, if they behave logically, will be anticipating some added advantage in moving. Also, potential movements from an origin to a final destination are likely to be influenced by **obstacles** at either source or destination, or on the journey. Such obstacles might include family pressures, national policy, travel costs, lack of capital, illiteracy, military service and language.

Lee was a sociologist and realised that the same feature might be perceived differently by different individuals — some might welcome the opportunity to live in a large city, with all the facilities it might offer, whereas others might find it cramped, polluted and depressing. In other words, it is more to do with what people *think* the world is like than any objective assessment.

Zelinsky suggested that there might be a transition to patterns of migration just as there is for demographic change. In his model, there are five stages:

(1) In a pre-industrial society, there is little residential migration, although there is nomadism.

(2) This is an early transitional stage of considerable rural–urban migration and the colonisation of new lands, with the associated growth of longer-distance migration (often in the form of emigration).

(3) In the third stage, rural–urban migration continues and there is a rapid rise in migration between cities.

(4) Rural–urban migration may continue, but at a markedly reduced rate; residential migration remains high, but in the form of migration in and between cities rather than emigration. There may be some immigration of unskilled workers, and highly trained professional workers may be exchanged between countries as a result of the operations of multinational companies.

(5) Advanced societies will have almost exclusively inter- or intra-urban migration, although new technology will reduce the need for migration and there will be less need for some types of circulation, such as long-distance journeys to work. Mobility between and within countries may be affected by state legislation.

Recent developments in migration theory have concentrated on distinguishing between personal motives for migration and global economic changes that produce forces for large-scale migration, for example the movement of people from Mexico into the 'sunbelt' states of the USA.

Migrations can be classified by motive, composition, distance, duration and direction

- *You should be able to categorise migration in terms of who migrates or why they migrate, and provide examples of where they go and how long they stay at their destination.*

Edexcel (A) Unit 2

The far-reaching social and economic effects of migration have led to a diverse range of classification systems.

Case studies

You need to know and be able to write about specific migration case studies across a range of scales, i.e. local through to global. The table below offers a range of possibilities.

Migration type*	Country	Migration event
A	Nigeria	In the late 1980s, up to a third of rural–urban migrants were forced, through lack of employment, to return to their villages from Lagos and other cities
	France	Counterurbanisation trends include Parisians moving to the 'rural' growth poles, such as Marne-La-Vallée (the site of EuroDisney)
B	France	Depopulation of the Massif Central through changing rural emphasis — migrants moving to Paris
	Zimbabwe	Rural poverty, agricultural mechanisation and land reform force people illegally into the city
C	Spain	Ex-patriot sunseekers retire to properties on the Spanish Costas
	Malaysia	Indonesians migrate to Malaysia to work in construction and agriculture
D	UK	Afro-Caribbeans came to work in the UK's factories after the Second World War
	Turkey	In-migration of guest workers into Germany after the Second World War
E	UK	Daily shop or visit to the newsagents; travelling sales representatives
	S. Sahel	Daily collection of firewood
F	Australia	Legal and illegal construction workers in Sydney's Olympic workforce
	Nepal	Ghurkhas recruited into the UK's army from Nepal
G	UK	Cheap breaks to European cities
	South Africa	Mozambicans working in South Africa's open cast mineral mines
H	Rwanda	Persecuted Tutsi Rwandans fled to Zaire in 1994; Zaire expelled refugees back to Rwanda in 1995
I	Kurdestan	Iraqi Kurds forced from their home area to Turkey after the Gulf War

* See Figure 15 (page 56)

Traditional migration classification is outlined in Figure 15. The letters in brackets (A–I) relate to the case studies in the table above.

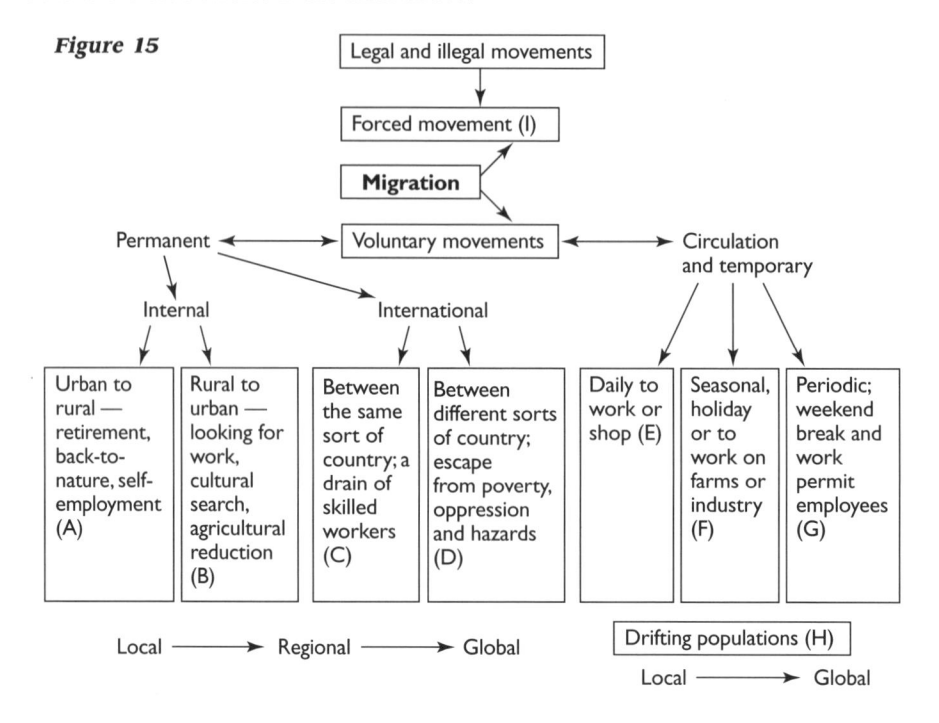

Figure 15

```
                        Legal and illegal movements
                                    ↓
                          Forced movement (I)
                                    ↑
                              Migration
                                  ↓↓↓
    Permanent ←————→ Voluntary movements ←————→ Circulation
         ↓                     ↓                  and temporary
     Internal             International
```

| Urban to rural — retirement, back-to-nature, self-employment (A) | Rural to urban — looking for work, cultural search, agricultural reduction (B) | Between the same sort of country; a drain of skilled workers (C) | Between different sorts of country; escape from poverty, oppression and hazards (D) | Daily to work or shop (E) | Seasonal, holiday or to work on farms or industry (F) | Periodic; weekend break and work permit employees (G) |

Local ——→ Regional ——→ Global

Drifting populations (H)

Local ——→ Global

The pace and extent of migrational change are emphasised by the following aspects of current/recent migrations. These changes in many ways supersede the traditional classification model.

Changes	Comments	Recent examples
Volume migration	Migrants are now able to move huge distances across the world	Europeans to the USA and Australia
Patterns of migration	Numbers of sending and receiving countries have increased	New senders include NICs, Russia, Romania and China; new receivers include Italy, Spain and the Gulf states
Migrant type	Forced migration is on the increase, as are smuggling and trafficking; voluntary (e.g. brain-drain) traffic has also increased	Albanians and Chinese to the UK; medical workers from the UK to the USA and Australia
Women > men	Emancipation of females in LEDCs has led to increasing numbers of movements	Philippino nurses to UK hospitals; Asian domestic help in Saudi Arabia
Globalisation	Improvements in communications and the global infrastructure have aided the migration process	Trans-Atlantic business travel; timeshare ownership on Portugal's Algarve

There is a variety of causes and constraints affecting people's decisions to migrate permanently

The influence on migration flows of push and pull factors
- *You should understand the differences between personal and family causes of migration and the larger forces that drive population movement (e.g. varying demand for labour).*
- *Your main focus should be on household decisions, but pay due attention to the constraints that limit movement as well as the positive forces of attraction.*
- *You should understand that people move for economic reasons but also that social and cultural reasons can be important.*
- *You need to understand the push and pull mechanisms.*

Motivations for migration

Not all forms of migration involve personal choice. There is a fundamental distinction to be made between **forced** migration and **voluntary** migration. Even this apparently simple distinction is not quite as straightforward as it seems. The Atlantic slave trade is an obvious example of forced migration, while retirement to the Costa del Sol is obviously voluntary, but the range between these is considerable. Refugees fleeing war have some sort of choice, as do the victims of natural disasters.

The reasons why individuals move are often very complex. A person may move for several reasons. For instance, individuals may move to support their families *and* to follow a lifestyle that they cannot follow in their own society. It is often difficult to establish why individuals *really* want to move. A person may want to leave a place to escape family pressure to marry, or to follow a certain career, but will find it difficult to admit this. It is useful to divide the reasons into two groups:
- the **push factors**, which lead people to leave one society
- the **pull factors**, which attract them to another society

What is known is that:
- in most cases, decisions to move involve groups of people (social/personal reasons)
- most move for/to work (economic reasons)
- many move to join relatives or to marry (personal/social reasons)
- an increasing number are fleeing persecution (political reasons)
- many are seeking the 'bright lights' of the city (socio-economic reasons)

More is known about the characteristics and nature of migrants:
- Most are young adults (20–35 years of age).
- Male and female movers are closely balanced — women moving short distances and men internationally.
- The two principle migrant groups are the highly educated and the poor, although not the very poor, since they cannot afford to migrate.

Economic, social, political and physical motives for migration over different distances and time scales
- *The driving force behind most, although not all, migratory movements is economic.*
- *The most important social motive is for marriage or to join relatives.*

- *Political motives are dominated by unrest, war and persecution (which cause refugees to flee a country or region) and by immigration policy, which may encourage or discourage migration.*
- *Physical factors are usually sudden events that cause people to flee. These include earthquakes and volcanic eruptions.*

The scale of human movement

It is useful to distinguish the scale or volume of migration, because different levels of migration can have profoundly different impacts on the societies that people leave and those to which they move. For instance, large-scale movements, like that of African slaves to the Americas, had a profound influence on both Africa and North and South America, as well as laying foundations for wealth accumulation in Europe.

Mass migration

This term is used to refer to large movements which occur over a short time. These movements often occur when large numbers of people seek to escape:
- natural disasters, such as floods, monsoons, hurricanes, earthquakes, volcanoes, diseases, famines and droughts
- social disasters, such as wars, ethnic/religious/political persecution and genocide, which is sometimes called 'ethnic cleansing'

These movements have huge social and economic costs in both the places which migrants leave and those in which they settle. Where these movements occur suddenly and unplanned for, they are often accompanied by major social disruption, as migrants:
- lose contact with their families and their places of origin
- are separated from their savings and resources and cannot provide for their needs
- suffer from poor conditions, which can affect their physical health
- suffer from fear and insecurity, which can affect their mental health

Individual migration and motivation

Motives of individuals who migrate may vary greatly. Some people migrate so they can:
- provide better social, economic and educational opportunities for their families. These individuals may continue to support their families from distant cities and countries through **remittance** payments
- escape their families and their expectations. These movements sometimes occur after arguments and conflicts, and lead people to sever ties with their families more or less completely and never return
- achieve short-term individual goals, such as saving money to build a house or start a business, and then return to their homes
- maintain beliefs that they are not allowed to hold or practise in their own societies or to follow lifestyles that they are unable to follow in their own society

Economic, environmental, social and demographic impacts of migration in countries of origin and destination

- *You should know at least two migrations to illustrate contrasting impacts on countries in different states of development. One should be an internal migration and one an international migration.*

- *The contemporary north–south drift of the British population could be used to illustrate several of these impacts at a national level, whereas Mexico–USA migration could provide an international example.*
- *The impact needs to be assessed in both positive and negative terms although, of course, there is no need to try to balance what might well be unbalanced effects.*

The impact of short-term (temporary) migration

This type of migration has more significant effects on both the community that they leave and the community in which they settle. The families of short-term migrants might receive more money but will be without a parent, or an adult child, who performed important social and economic duties for the family. Thus, children left behind may find themselves without a parent for important parts of their lives. The **benefits** of short-term migration to the family may be offset by certain **costs**.

Short-term migration also has an impact on the societies in which migrants settle. These migrants earn money, which they spend on living expenses such as food, accommodation and transport, as well as services such as insurance, banking and health. They may buy tools, clothes, food, appliances and equipment to send home. All of these purchases create demand in the economy in which they settle which can result in economic growth. They also pay taxes on their incomes in the countries in which they work. In many states, they pay indirect taxes, such as VAT, on everything that they buy. This includes both goods such as food and cars, and services such as rent and banking transactions. For these reasons, short-term migrants may produce economic growth in the places where they settle. At the same time, because they return to their country of origin, temporary migrants do not incur the high costs of supporting them in their retirement.

The impact of permanent migration

The impact of long-term migration for the sending society is often serious. Where large numbers of people leave for the longer term, the donor communities lose:
- their labour and skills — 'human capital' or 'social capital'
- the income that the migrants produced
- the food, goods and services that the migrants produced
- the social contributions to their families and communities

While the donor community gains from money, tools, gifts and food, the value of these may decrease over time. Migrants may send significant amounts to support their families and communities while they are single. As they marry, buy homes and establish families, they face increasing expenses in their new homes. It often becomes difficult for them to maintain their contributions to their donor communities. Where this happens, the donor community may lose the migrants' economic and social contributions.

The impact of long-term migration for the receiving society is also more significant. The receiving society experiences some economic growth as the migrants join the workforce. The migrants pay income tax and expenditure taxes such as VAT, which expands the receiving government's revenues and allows it to increase services. The migrants also spend some of their income on food, homes, clothing, education and recreation. Their expenditure generates demand for the goods and services which are provided by private companies. This creates economic growth, leading to more jobs.

On the other hand, the receiving society faces the costs of providing health, education and welfare services for migrants' families which are unable to work. This usually happens:
- during economic recessions, where businesses or factories close and migrants and their families are unable to find work
- where the labour market changes, so that certain posts which migrants have traditionally filled disappear and are replaced with new ones which migrants do not have the skills for
- after retirement from the workforce

By separating and understanding the various types of impacts that result from migration, it is possible to assess the importance of migration in the short, medium and longer terms for the societies which migrants leave and those in which they settle. To do this type of cost–benefit analysis, various consequences have to be distinguished and measured.

Demographic consequences
Migration may have profound effects on the size, structure and growth patterns of populations. It affects both the populations of the places that people leave and the populations of those in which they settle. These effects vary with different types of migration, the age structure of the migrants and the length of migrants' stays. The absence of large numbers of either men or women may have a limited impact on the sending society in the short term, but if they are absent for longer periods of time, their absence will have significant effects on population growth rates in the medium and longer terms.

Social consequences
Social consequences include the impact on individuals, families and communities. They can be either positive or negative and they can change over time. Disruption of family life is probably the most significant consequence of modern population movement.

Economic consequences
Migration almost always has very significant effects on economies. These effects vary with different types of migration, the skills of the migrants and the lengths of time involved. Most observers believe that migration has a largely positive impact on the economy of receiving countries, which is hardly surprising, since the demand for labour is the most important driving force behind most major migrations.

Political consequences
Migration can have an impact on politics in both the places which people leave and those to which they move. Governments might have to make policies to attract migrants, to persuade migrants to return, or to limit migration and ensure that they

have access to the skills that they need. These political effects vary with different types of migration.

Social and cultural opportunities and challenges at local and national levels
- *You could study this with reference to the UK or another country, and in a particular local area within that country with a significant number of immigrants.*
- *You should be able to present positive and negative aspects of migration, including racism and the impact of racial tension on the geography of ethnic communities, cultural enrichment through diversity and the exchange of ideas.*
- *The challenges are a consequence of the frequently negative attitudes faced by immigrants, and the difficulties they have in integrating with local communities.*
- *The opportunities are the potential social and cultural benefits that can be enjoyed by both migrant and host communities.*
- *In many cases, the economic benefits are obvious to both communities, except during periods of economic recession.*

The reception and treatment of migrants

The reception that migrants encounter in the receiving society will depend on a number of factors. Migration often brings together people with very different worldviews and lifestyles, so it is not surprising that there is often some suspicion on the part of both the hosts and the migrants. Where the worldviews and lifestyles are very similar, there are typically lower levels of suspicion, as with migration within Europe. Where groups do not understand very much of each other's worldviews and lifestyles, there is typically:

- a greater dependence on social stereotypes (that is, generalised views of others based on a very limited personal contact and very few observations)
- a higher level of prejudice, based on social stereotypes rather than personal experience
- a higher probability of ethnic discrimination (that is, where a person's beliefs about a migrant group influence their behaviour towards that group and lead them to treat members of that group differently from others). While this discrimination occasionally leads to more favourable treatment of members, it usually results in less favourable treatment.

Social stereotypes, prejudice and discrimination often combine to produce 'self-fulfilling prophecies', which seem to 'confirm' social stereotypes and to justify prejudice and discrimination.

But members of the host society are not alone in this conduct. Migrants also discriminate against their own people. For example, experienced migrants may take advantage of new arrivals' lack of knowledge to discriminate against them. This is well documented in cities in the USA (e.g. loan-sharking).

Some migrants are more vulnerable than others to prejudice and discrimination. Those migrants who are in a country legally can openly claim and pursue all of the rights of citizens or permanent residents. They can seek assistance from lawyers, government departments and trade unions to pursue those people or companies whom they believe have discriminated against them in any way. They can pursue their rights without fear of being removed from the country and losing access to income that they need to support their families.

Those migrants who are in a country illegally are much more vulnerable to discrimination and are much more likely to be denied legal and human rights. If they pursue their rights, they risk being identified and removed from the country and losing the income they need to support families.

Positive opportunities also exist for migrant communities and for receiving societies. The benefits include:
- cultural enrichment, such as the development of musical styles through the blending of different traditions
- the contributions of migrant communities to broader popular culture (local and national sporting teams are an obvious example)
- social enrichment, with inter-marriage gradually dissolving prejudice and racism

The impact of migration on the physical environment
- *You should understand that the impact might be positive or negative.*
- *You should know that the impact will be on both the donor country and the recipient country.*

This could be studied at a local scale, with the debate about greenfield house building in the southeast of England and the desertion of central urban areas by the more mobile middle classes in many conurbations. In this case, migration has caused changes in population density, with inevitable impacts on the physical environment.

Mass movements, such as the migration to the Amazonian territories from southern Brazil, have been instrumental in the destruction of the rainforest through burning to make way for agriculture.

Abandoned areas also change. The natural vegetation changes and climax communities can re-establish.

Desertification is often blamed on the in-migration of farmers, who exploit soils to such an extent that the nutrient levels decline, plants cannot establish and thus surface runoff increases, causing rill and gully erosion and major landscape changes. The changes that have taken palce in old mining areas might also be useful to illustrate the idea. In south Wales, the quality of river water, the infiltration rate, river discharge and erosion rates have all been affected by the closure of mines and the out-migration of the population.

A general link back to the idea of exceeding carrying capacity is appropriate here.

Questions & Answers

This section contains six Unit 2-style questions. You should note that here there is only one full-length question specifically for each of these sections outlined in the content guidance (Questions 1, 2 and 3). In the unit test there will, of course, be a choice of two. Questions 4, 5 and 6 are extended-writing type questions; try to answer them all. They attempt to give you further extended writing practice, and to cover more of the specification.

Two model answers are provided after each question. The first is of a typical grade-C standard (candidate A) and the second is of a good grade-A standard (candidate B).

Examiner's comments
The answers are interspersed by examiner's comments (preceded by the icon *e*). These show where credit is due and where errors have been made or improvements are needed. They highlight problems such as irrelevancy, lack of clarity, lack of focus on the wording of the question or shortage of case study detail.

The comments indicate how each answer would have been marked in an exam. In questions only worth a few marks, the examiners may be instructed to give a mark for each relevant point that the candidate makes (up to the maximum). In higher-mark questions, there may be a mark or two for a 'basic' answer (such as the naming of an appropriate process), with additional marks available for 'development' (describing the mechanism of the process). In the final part of the question (usually worth 6 marks), the examiners use a system of 'bands'. These are defined as follows:

Level 1: 1–2 marks
The answer offers simple ideas or a list of points, not all of which may be relevant. A located example (if requested) might not be given.

Level 2: 3–4 marks
Some knowledge and understanding is evident, but cause–effect links may be unclear or simply stated. There may be a lack of range of ideas/information should this be required. Located detail may be unconvincing.

Level 3: 5–6 marks
The answer reveals accurate knowledge and clear understanding. Explanations are explicitly stated with cause–effect ideas well argued. Terminology is used accurately and located detail provided.

Question 1

Population characteristics (I)

Study the graph below, which shows crude birth rates and death rates by continents.

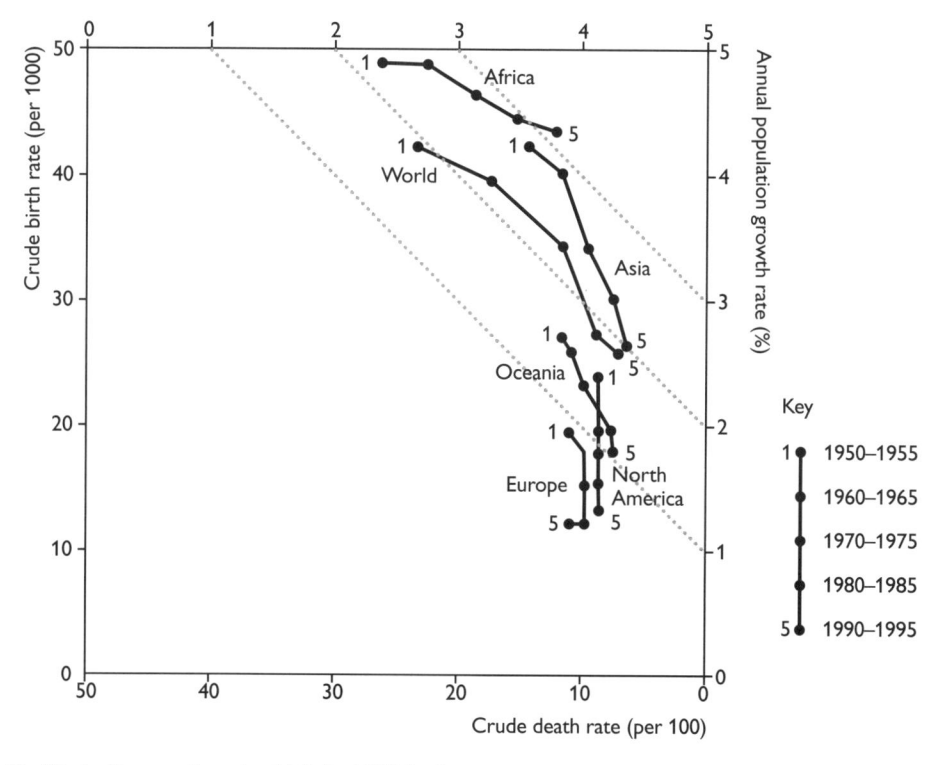

[Graph: Crude birth rate (per 1000) on left y-axis, Annual population growth rate (%) on right y-axis, Crude death rate (per 100) on x-axis. Shows data points for Africa, World, Asia, Oceania, Europe, and North America across time periods. Key: 1 = 1950–1955, 2 = 1960–1965, 3 = 1970–1975, 4 = 1980–1985, 5 = 1990–1995.]

(a) (i) State the continent which in 1995 had:
    (1) the lowest crude birth rate
    (2) the lowest crude death rate (2 marks)
  (ii) Suggest reasons for the answers given in (a)(i). (6 marks)
  (iii) Describe and give possible reasons for changes over time in North America and Africa. (4 marks)
(b) (i) How is annual population growth rate calculated? (2 marks)
  (ii) With reference to located examples, outline the economic and social effects of high and low population growth? (6 marks)

Total: 20 marks

Answer to question 1: candidate A

(a) (i) (1) Europe; (2) Asia

> *e* Correct, for 2 marks. The ability to read and interpret graphs is an extremely important skill and therefore one that needs practising.

AS Geography

question 1

(ii) (1) Europe is mostly in the later stages of demographic transition, with smaller families resulting from a lower need for child labour, with female emancipation along with readily available contraception, and with family planning advice.

e Words like emancipation are fine, but its impact needs to be made clear. Greater play could have been made of children being a cost rather than an asset. This answer is worth **2** out of 3 marks.

(2) Asia has, in recent years, become more developed, with greater access to medical facilities, a better standard of living and improved sanitation. All these factors have led to a lesser chance of disease spreading, and a longer life expectancy.

e 'Asia has in recent (?) years become more developed' is not a direct answer to the question and is therefore unnecessary. More could be said here; comments on education, diet and, above all, the low median age of the population would have received credit. This answer gains **1** out of 3 marks.

(iii) As can be seen from the graph, North America's death rate has remained the same since 1950, with only the birth rate dropping as the third stage of the demographic transition is completed. In contrast, Africa's birth rate has fallen little in comparison to its death rate, with the population growing as it completes the early expanding stage of the demographic transition model.

e 'As can be seen from the graph' is not necessary; we know we are dealing with a graph! However, what is said is accurate if a little simplistic: straightforward description with no attempt to explain. This is awarded **3** marks.

(b) (i) The annual growth rate is calculated by finding the difference between the death rate and the birth rate, and converting this into either a positive or a negative percentage compared with the overall population total.

e This answer scores **1** mark. The clearest way to answer this is to state the equation:

$$\frac{\text{net annual BR}/1000}{\text{net annual DR}/1000}$$

(ii) In Bangladesh, the high birth rate and falling death rate have led to massive population growth in the recent past, with over-population becoming a major problem. Due to the country's inability to generate large crop yields, the GDP of the country has fallen to just $220, with many families struggling to survive. The standard of living is poor, with a very low calorie intake. Disease often has a cruel impact upon the vulnerable population. The high population growth has also meant that the country is finding it very difficult to develop, with the birth rate continuing at a high level.

e This answer is written in general terms and low population growth is not addressed. The location is recognisable, and what is written is sound, but once

Edexcel (A) Unit 2

again this candidate has not read the question properly and has only addressed high population growth. This is awarded 3 out of 6 marks.

■ ■ ■

Answer to question 1: candidate B

(a) (i) (1) Europe; (2) Asia

e Correct, for 2 marks.

(ii) (1) Europe is made up primarily of MEDC countries, with high education standards and strong economies. With more and more people concentrating on careers and a wide availability of family planning facilities, the BR in Europe is kept reasonably low. Also, people in Europe have, in recent years, been having fewer children and getting married later, which keeps the BR low.

e A number of relevant issues are put forward here, for 3 marks.

(2) Asia has the lowest crude DR, due to the high medical technology available in countries such as Malaysia and Japan. These countries have profited from the boom in high technology goods, and so there is money available to have an excellent quality of life and superb healthcare facilities, thus prolonging life.

e This answer covers some of the social and economic factors. The general focus is recognisable, but it lacks sufficient development of the factors/areas identified for full marks. It is awarded 2 out of 3 marks.

(iii) Africa has had a significant drop in DR due to improving medical facilities, vaccines and a better quality of life being available. BR has dropped slightly due to increased contraception and education. North America has had no change in DR because quality of life and medical care have been constantly good, but it has experienced a large drop in BR due to contraception and modern lifestyles in MEDCs.

e The basics are here and so the answer just about deserves full marks. More comment on the past 45 years and something more than straightforward description would make for a more solid answer.

(b) (i) Annual population growth is calculated by subtracting the crude DR from the crude BR, leaving the annual growth rate.

e It is so important to know and understand basic definitions. This good candidate has missed out on 2 straightforward marks.

(ii) In New Zealand, there have been many economic and social implications arising from high and low population growth. In the 1980s and early 1990s, New Zealand was experiencing low population growth and was therefore underpopulated. This meant that the economy began to dip due to the vacancies in the employment market. However, immigration and higher BR caused a growth

question 1

in population, which boosted employment and the economy, bringing productivity up. However, many New Zealanders were upset that the government should resort to allowing in immigrants in order to fill vacant positions.

e The first sentence is not necessary. The candidate is aware at a basic level of the problems of under-population. The obvious imbalance in the answer locates it firmly at Level 2. It is coherent, it uses geographical terminology and it is located — 5 marks are awarded.

Population characteristics (II)

Study the diagram below, which shows a Malthusian view of the relationship between population growth and food production.

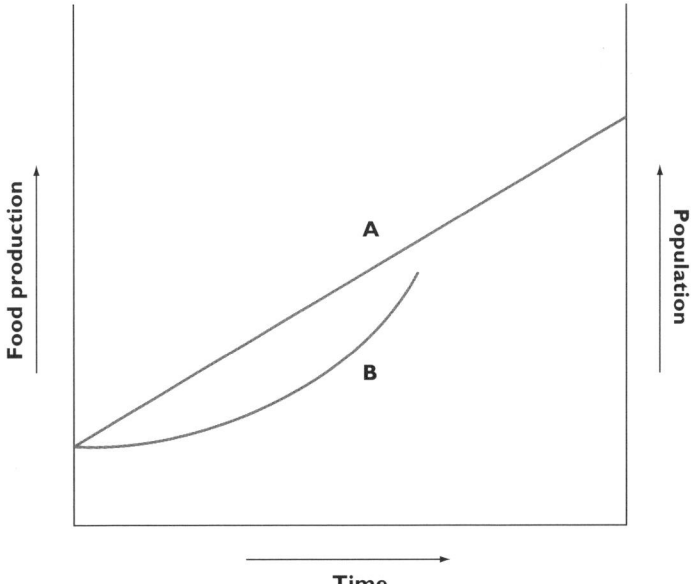

[Diagram: graph with Food production on left y-axis, Population on right y-axis, Time on x-axis. Line A rises linearly; line B curves upward exponentially below A.]

(a) (i) Which of the lines labelled **A** and **B** shows:
   (1) population growth?
   (2) food production? (2 marks)
(ii) Describe and explain how population growth might change up to the end of the time period. (3 marks)
(iii) Define the term *underpopulation*. (2 marks)
(b) With reference to the diagram, suggest reasons why:
   (i) food production might increase more rapidly than shown (4 marks)
   (ii) the rate of population growth may be less than shown (4 marks)
(c) With reference to located examples, describe the characteristics of an underpopulated region or country. (5 marks)

Total: 20 marks

■ ■ ■

Answer to question 2: Candidate A

(a) (i) (1) B; (2) A

e Correct, for 1 mark.

question 2

(ii) The population might increase rapidly until nearly the end of the time period shown, before falling back again. This is because people will go on having children even though there are not enough resources for them. It will take some time for them to realise that they are having too many children for them to survive.

e The basic description is good, for 1 mark, but it does not really get to the point of explaining the fall described. The relationship between food supply and population is not touched on.

(iii) Underpopulation occurs when there are not enough people in a country.

e Not enough for what? The idea of a deficiency is here, for 1 mark, but it is not developed.

(b) (i) Malthus was wrong because he did not foresee the agricultural revolution ('green revolution'). This happened after he wrote his book and meant that much more food was produced in many countries, meaning that the famine that he predicted never happened. The same thing has happened in India, with new types of seeds and much better farming methods.

e This only gets to the point in the last sentence with the 'new types of seeds'. The confused history doesn't help, but what is missing is a fuller account of the techniques or 'methods'. If this candidate had outlined some of these methods, they would have been rewarded, despite the confused chronology. This is awarded 2 marks for the basic idea of a 'revolution' in food output using new seeds.

(ii) In the last 100 years, we have learnt how to control family size with modern birth control methods. Family planning advice is available and there is no need to get pregnant accidentally. As a result, population has not increased in MEDCs as fast as Malthus thought. In LEDCs, they don't have such good methods and there is still a problem. Improved medicine has also led to a fall in the number of infant mortalities.

e There is a very narrow focus here: firstly, on birth rate and then on family planning. If this answer had been developed by the use of examples (Mauritius?), it would have gained more credit. The last line suggests real confusion, in that a decrease in infant mortality would have a *positive* impact on population growth rate and not a negative one. 1 mark is awarded.

(c) Countries with not many people may have all sorts of problems. They are probably quite harsh, climatically speaking, which explains the numbers, but there may be resources, such as oil in Alaska, which need people to exploit them. These people might have to be paid a great deal to work in harsh areas and the country will probably have to recruit them from abroad. If there are not enough people, the government might have a pro-natalist policy, like France, or encourage people to come (like Canada). There might be real problems with numbers of skilled people like doctors, so high wages will have to be paid.

Edexcel (A) Unit 2

🖉 This is a muddled response of disconnected ideas, with some confusion between sparse, low-density areas and underpopulation. However, a number of valid points are made (almost by accident with the 'Alaskan' example) and it is awarded 4 marks.

■ ■ ■

Answer to question 2: Candidate B

(a) (i) (1) B; (2) A

🖉 Correct, for 1 mark.

(ii) Population is likely to rise above the food production line before falling rapidly back below that line. Because a population increases geometrically, the population growth is likely to be more rapid than food production. However, when the population exceeds the food supply, the full population may not be supported, i.e. the carrying capacity will have been exceeded. In response to famine (natural check), the population then decreases to a level where it is more in balance with the food supply.

🖉 It is not quite an explanation, but the basic Malthusian idea is presented clearly here, with an accurate use of terminology (such as carrying capacity), for 3 marks.

(iii) There are not enough people to exploit fully the available resource base. Therefore, an increase in population would lead to an increase in economic prosperity.

🖉 This is a very good answer, for full marks.

(b) (i) This may occur in response to developments in technology and farming techniques. For example, development of high-yield crops or use of multi-cropping will increase the output per unit area of land. Also, mechanisation of the industry may result in the replacement of people with more efficient techniques of planting and harvesting, while the area of land used for crop production may be increased. Therefore, line A would curve upwards — i.e. exponential growth.

🖉 The description of intensification and increased yield is sound and this candidate adds the idea of increasing land area devoted to crops. This is a thorough response, for 4 marks.

(iii) Population growth is composed of natural increase and immigration on a global scale. Birth rate (fertility) may fall below the replacement ratio, for example as a result of a compulsory education act or women marrying later. This would also be encouraged by demographic change, for example an ageing population. Also, the death rate may increase in response to the spread of disease or due to an ageing population (a large proportion over 75). As a response, the B line may continue to follow a straight path rate rather than curve upwards.

71

AS Geography

question 2

> *e* There is some confusion over terminology here (migration not immigration), but the answer is thorough in that it offers reasons for changes in both birth and death rates that are appropriate to a continuation of the trend shown. Full marks are awarded.

(c) In an underpopulated region, there tends to be vast areas of land that are uninhabited or have a fairly sparse population, for example the rural Scottish Highlands. Also, there may be resources that are yet to be fully developed and explored, which represent a potential capacity for increased population. There tends to be the desire for the population to increase, as it is recognised that this is likely to improve living standards. In Canada, for example, there is an 'open-door' policy whereby immigration is very much encouraged to take place. This is based around government action and schemes.

> *e* This is a Level 2 response and is awarded 3 marks. The candidate has offered sparseness, resource abundance and immigration policy, but labour shortages and wage rates are missing and the examples are not developed beyond naming two places, one of which (the Scottish Highlands) is not obviously underpopulated.

Question 3

Settlement patterns (1)

Study the map and cartoon below. They illustrate the character and location of spontaneous settlements (shanty towns) in a large city in an **LEDC**.

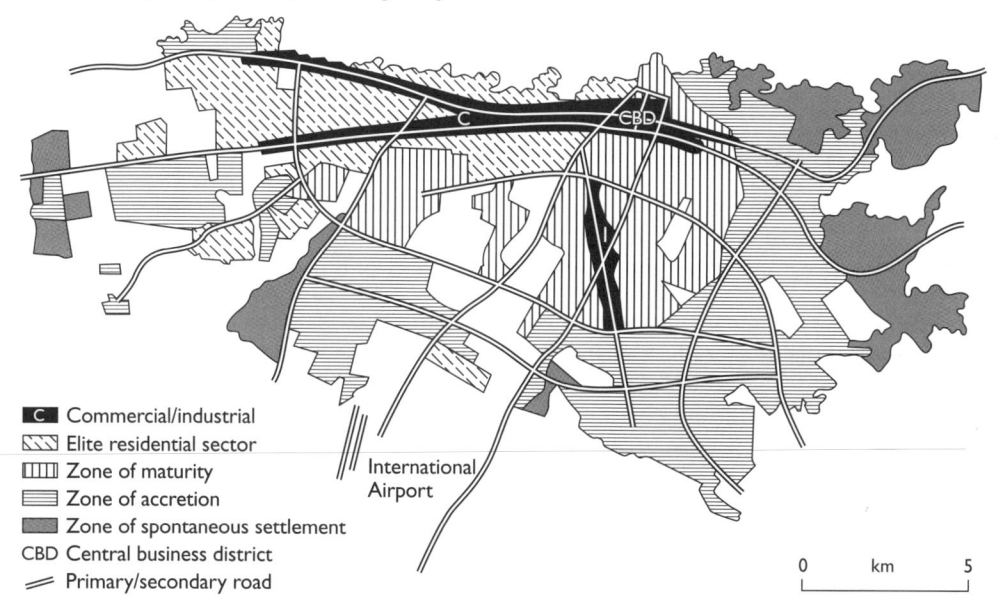

- **C** Commercial/industrial
- Elite residential sector
- Zone of maturity
- Zone of accretion
- Zone of spontaneous settlement
- CBD Central business district
- Primary/secondary road

'What slums?'

(a) (i) Explain what is meant by the term 'spontaneous settlement'. (2 marks)
    (ii) Describe the distribution of spontaneous settlement shown on the map. (3 marks)
    (iii) Suggest reasons for this distribution. (3 marks)

AS Geography

question 3

(b) (i) Many primate cities in **LEDCs** have very extensive areas of spontaneous settlement. Suggest reasons for this. (3 marks)

(ii) What does the cartoon suggest about the character of **CBD**s of large cities in **LEDC**s? (3 marks)

(c) With the use of located examples, outline the issues associated with an edge-of-town retail development? (6 marks)

Total: 20 marks

Answer to question 3: candidate A

(a) (i) Spontaneous settlements originate from a single shanty house. Growth from this through other buildings nearby results in a spontaneous settlement, where people settle near others, for example in areas of Bogota.

e This answer is only partially relevant. The illegality of these settlements and speed with which they are built are not covered, although there is some grasp of their growth. 1 mark is awarded.

(ii) Spontaneous settlement occurs primarily towards the remote east or west of the area, with most being situated either on or near a main road. Most of the settlements are grouped closely together in areas of this city, although there are several smaller spontaneous settlements to the north, some distance away from the larger settlements.

e The candidate identifies locations and describes the sites of spontaneous settlements, for full marks.

(iii) There is obviously a physical barrier at the northern end of this city which affects distribution of the shanty towns to the north. Most settlements are near a major road as this offers easy access and provides the possibility of gaining money from commuters entering or leaving the CBD. They are closely clustered in the main, in order to keep a sense of community and because the derelict areas needed to build the settlements would have been in close proximity to each other.

e The answer identifies the need for vacant land, but fails to use the geography of the subject — the concept of bid rent, for instance. It also identifies opportunities for roadside dwellers, but not why they originally located there. 2 marks are awarded.

(b) (i) As primate cities in LEDCs have large numbers of people compared with the accommodation space available, people with no homes have to make their own settlements on any spare land. There is not the money or technology in LEDCs to house the entire population and so it is often easier for both the people and the government to allow areas of spontaneous settlement to take place.

Edexcel (A) Unit 2

e The attractiveness of the primate city is not very clearly identified. The lack of money is included and understood as a handicap. The answer also recognises that governments willingly overlook spontaneous settlement development. 2 marks are awarded.

(ii) The cartoon suggests that those who work in the CBDs of large cities in LEDCs are not aware of the problems that exist in the settlements outside the CBD and are more concerned with making money and living their own lives than helping those in slums. It illustrates ignorance of the problems faced in LEDC cities by those who are at the top.

e There is a simple grasp here, although the answer is highly generalised. There is little comment about the contrasts, but some useful remarks about the attitudes are included for 2 marks.

(c) Out-of-town shopping centres have developed in many MEDCs in recent years. One of the largest in Britain is Bluewater, near London. These centres are controversial, with many issues raised both for and against them. The biggest problem with them is that they have a negative impact on nearby shopping centres in towns and cities. If people go to the large out-of-town centres, they do not use the CBDs of traditional towns. As a result, less money is spent in these places and they decline over time, with only charity shops and building societies being left.

e There is some hint of understanding here, but a long time is spent saying little — 'many issues raised both for and against them' is typically unspecific and not developed later on. We never discover the advantages. The only impact covered is the effect on nearby towns, but this is not developed and nor is it located. This deserves just 1 mark.

■ ■ ■

Answer to question 3: candidate B

(a) (i) A spontaneous settlement is a housing area devoid of the most basic amenities, with buildings that are unplanned, unregulated and mostly of poor quality. They are often house squatters.

e This mentions the lack of amenities and implies the illegality (squatters) and speed with which the settlement is built, for 2 marks.

(ii) The settlements are close to the zones of accretion and are often abutting main routeways or industrial areas. The settlements are also located on the periphery of the settlement, in the areas furthest from the CBD and the elite residential sector.

e The candidate has identified the locations of spontaneous settlements and has used the resource sensibly. Some compass direction would have supported the description. All the same, this answer is worth 3 marks.

AS Geography

question 3

(iii) The settlements are found on the periphery, as this is often the cheapest and least desirable land, not wanted by those who can afford to build better homes. The settlements often cluster around areas that have services, in the hope that further resources and services can be gained.

This alludes to bid rent with the idea of 'cheap' land. The candidate could have expanded on the peripheral nature of spontaneous settlements, being the first places at which rural–urban migrants arrive. 2 marks are awarded.

(b) (i) Although these cities offer more job opportunities than other places in LEDCs, they cannot support the numbers of migrants that flock to them (over 1000 a day into Mexico city), often due to the mechanisation of agriculture. With no cheap housing available, the migrants are left to build their own, with the governments not having the resources to allow them to offer the migrants appropriate aid.

This offers economic reasons for the move to the city and the building of spontaneous housing. There is also the realisation that migrants cannot be supported by the state. This answer is worth 3 marks out of 3.

(ii) The cartoon suggests that there is a great contrast between the often attractive CBD of an LEDC and the surrounding slums and their impoverished inhabitants. The CBD often has luxury apartments and skyscrapers, with many TNCs creating bases in the district.

Some of the features of the LEDC CBD are outlined, for 1 mark, though its size, location and attitudes are not considered.

(c) Out-of-town retail developments have several impacts, both direct and indirect. In the case of Merry Hill, near Dudley, there were issues raised in the planning stage with the use of agricultural land and the loss of habitat. Since its opening, the loss of trade in nearby Dudley has been significant, with several shops closing, especially food stores. This can lead to a spiral of decline. The increase in traffic generated by Merry Hill has impacted on local residents, who have also complained about noise and air pollution. Some have benefited by gaining employment there and some argue that even Dudley will benefit eventually, as more and more people are pulled into the area from the wide catchment area of the centre.

This is a good answer, at Level 3, for 5 marks. It has a range of issues, both negative and positive, and tries to address direct and indirect effects. The absence of much location (i.e material specific to Merry Hill rather than any other edge-of-town development) prevents it from being awarded full marks.

question 4

Settlement patterns (II)

Study the diagram below, which shows a model of land use in **North American cities**.

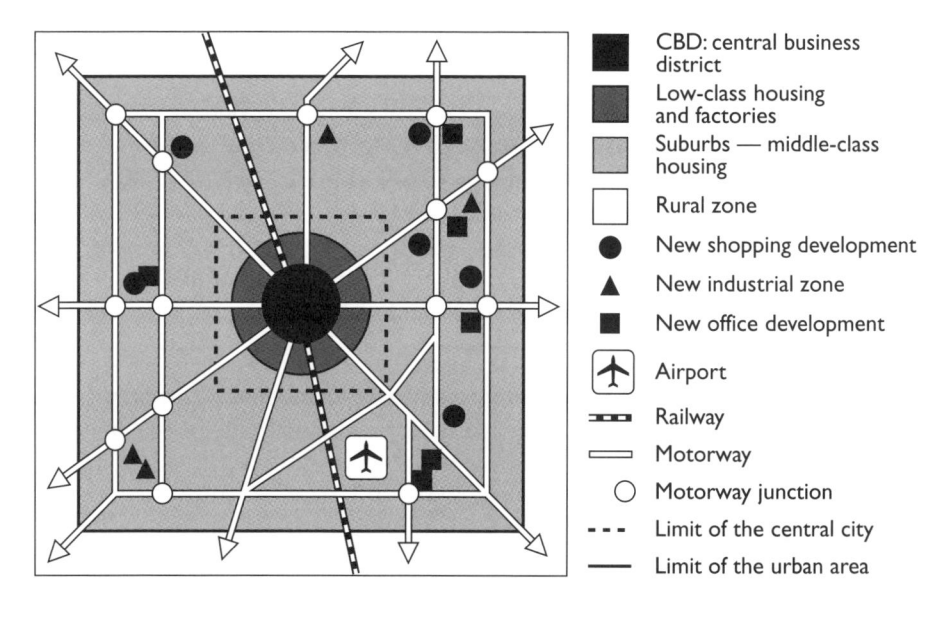

- ■ CBD: central business district
- ■ Low-class housing and factories
- ▨ Suburbs — middle-class housing
- □ Rural zone
- ● New shopping development
- ▲ New industrial zone
- ■ New office development
- ✈ Airport
- ▭▭▭ Railway
- ▭▭ Motorway
- ○ Motorway junction
- - - - Limit of the central city
- ─── Limit of the urban area

(a) (i) **Describe the distribution of the new industrial zones.** (3 marks)
   (ii) **Suggest reasons for this distribution.** (3 marks)
   (iii) **Outline two limitations of urban land use models, such as that shown in the diagram.** (4 marks)
(b) (i) **Define the term** *suburbanisation*. (1 mark)
   (ii) **Identify the impacts of suburbanisation on rural areas.** (3 marks)
(c) **With reference to located examples, outline recent changes in the inner city.** (6 marks)

Total: 20 marks

■ ■ ■

Answer to question 4: candidate A

(a) (i) They are all located next to the motorway (most are near a motorway junction). The distribution is uneven, as most developments are situated in the east. All the developments are in the suburbs.

    e This is very brief, but it has the main points and has addressed the pattern directly ('uneven'), for 2 marks. However, it is wrong to suggest that they are 'in the east'.

  (ii) All the developments are situated near to the motorway and its junctions for ease of access to both workers and customers. They are all situated in the middle-class suburbs, as the land is probably cheaper (bid rent theory).

AS Geography

question 4

e There are three ideas here: access to customers (likely); access to workers (likely); and land prices (very possible). The idea of the supply of components is not here, but two developed ideas get the full 3 marks.

(iii) They only show a typical town/city. All towns/cities are different. The profit of a company will determine where it situates itself within a town (bid rent theory).

e The second comment makes no sense in terms of the validity of models. The first is better ('typical'), but although it is unarguable that towns and cities vary, the point of models is that they allow the key features to be identified. 1 mark is awarded for the basic idea that models simplify reality.

(b) (i) An area which is developing into a fully urbanised area.

e This suggests that suburbanisation is a stage in the development of urban areas, which is not the case. No mark is awarded.

(ii) A suburb could be described as a transitional stage between an urban area and a rural area. The impact of a suburb on a rural area could be the influence of urbanisation — people moving from towns to rural areas. This might destroy rural habitats and use up valuable farmland.

e There is continued confusion, although two points are made at the end. The destruction of rural habitats is a very general remark, but the use of valuable farmland is more specific. It is awarded 1 out of 3 marks.

(c) Bath has experienced recent changes. It used to be a spa town, with a small high street. Over the past few years, Bath's high street, Milsom Street, has expanded rapidly, with an increase in the size and quality of the shops locating there. The main high street is dominated by national chain stores and some highly expensive designer outlets, which reflect the fact that Bath has become a very important tourist destination in recent years. As a result, more restaurants and tourist-oriented shops have opened; this is to be expected, as bid rent theory suggests that they will be able to afford more central sites.

e This answer takes a very narrow focus on 'inner city'. Only retail functions are addressed and while what is here is quite capably done, the omission of any other type of land use or any other type of change restricts it to Level 2. It is a shame that the candidate did not take a wider view of the question. It is awarded 3 marks out of a possible 6.

■ ■ ■

Answer to question 4: Candidate B

(a) (i) There are four new industrial zones, each located next to a motorway junction. Two are in the far southwest of the city, while the other two are in the northeast. The southwestern pair are close together.

Edexcel (A) Unit 2

e This is a comprehensive answer, which describes the distribution well, for 3 marks.

(ii) Access is probably the key. These industries are close to motorways, allowing good distribution of their products. They are on the margins of the cities, perhaps because of planning regulations keeping industry away from central city areas and middle-class residents who would resist industrial development.

e There is good geography here and although there are only two ideas, the mark scheme allows for full marks if these ideas are developed, as they are here.

(iii) Models are generalisations and this obviously simplifies complex patterns. For example, it is obvious that land use is often more mixed than models suggest, with retail, industrial and residential use being found together, even though one of them might be dominant.
    Cities change over time and so a model may become outdated. This happened with the model of Chicago drawn up by Burgess, when black immigration slowed down.

e This a full answer. Both ideas are legitimate and have been developed and illustrated, for full marks.

(b) (i) An area which is growing through the in-migration of people from inner-city areas — the less densely built-up outer residential zones.

e This is a full answer, covering two aspects of the idea.

(ii) Suburbs use up valuable farmland and their growth can destroy valuable ecosystems such as wetlands, hedgerows and woodland. They also transform rural communities as incomers settle in villages which become suburbanised and lose their identity.

e There are some good ideas here about the loss of land and habitat, though the social impact is identified but not developed. We are not told what 'losing their identity' actually means. This is awarded 2 marks out of 3.

(c) Bath has experienced many changes in recent years. The rapid growth of inner-city shopping facilities reflects the growth of tourist functions. This has changed the nature of the shopping centre, with designer outlets and smart restaurants opening. Bath has never developed a distinctive office and commercial core, but these functions are distributed on the margins of the retail zone, especially in the west on the Bristol Road. Partly as a result of its popularity with tourists and day visitors, there have been particular problems managing traffic in the city centre, with many changes in one-way systems. A number of derelict sites close to the Abbey have been adopted for car-parking.

e This answer takes a wide focus on the 'inner city', with retail, commercial and transport functions addressed. There are locations (Abbey, Bristol Road) which identify the city, and some specific knowledge is well linked to the title. This is a Level 3 answer, worth 6 marks.

AS Geography

Question 5

Population movements (I)

Study the diagram below, which is a model of international migration, and answer the questions that follow.

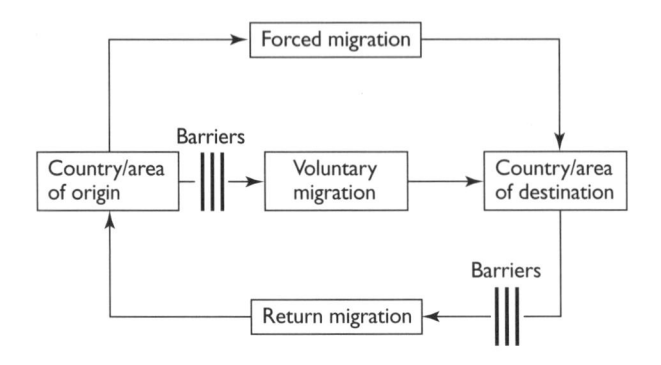

```
                    ┌──────────────────┐
         ┌─────────▶│ Forced migration │─────────┐
         │          └──────────────────┘         │
         │                                        ▼
┌─────────────┐  Barriers  ┌───────────┐   ┌──────────────┐
│ Country/area│    │││    │ Voluntary │   │ Country/area │
│  of origin  │───▶│││───▶│ migration │──▶│ of destination│
└─────────────┘            └───────────┘   └──────────────┘
         ▲                                        │
         │                              Barriers  │
         │          ┌──────────────────┐   │││   │
         └──────────│ Return migration │◀──│││───┘
                    └──────────────────┘
```

(a) (i) Define the term *forced migration*. (2 marks)
    (ii) How and why might 'voluntary' and 'forced' migrations differ in terms of:
        (1) their population composition
        (2) the country destination (6 marks)
(b) Why might:
    (i) internal migration be encouraged? (3 marks)
    (ii) international migration be discouraged? (3 marks)
(c) With reference to examples, show how physical factors might influence the volume of migration. (6 marks)

Total: 20 marks

■ ■ ■

Answer to question 5: candidate A

(a) (i) Forced migration is when a person or group of people are forced to move or migrate due to factors such as war. They are impelled to leave.

    e This has the compulsion idea, and the example of war, for 2 marks.

(ii) (1) Voluntary migration usually consists of people seeking work, such as young males, and people searching for a better life (also males). Forced migration consists of a great variety of people, including women and children.

    e This addresses 'how', but not 'why'. The command words matter! 1 mark is awarded.

(2) Voluntary migration is usually from a country of great poverty — usually an LEDC, since these people are searching for a better life. Forced migration can be from any country (MEDC or LEDC) experiencing a forcing factor, such as war or a volcanic eruption.

80

Edexcel (A) Unit 2

> *e* The origins are covered, but not the destinations! Destinations were asked for here, so no marks are awarded.

(b) (i) Governments may be trying to move people around a country because of a natural disaster, having to evacuate people from one area. This happened in Florida with Hurricane Andrew. Sometimes, they might try to move people away from areas that are too crowded or to areas that are underpopulated. The Russians did this with Siberia.

> *e* The first idea is limited and not, strictly speaking, a migration. The second idea is far too vague to gain credit, but the idea of populating underpopulated regions is legitimate and illustrated, for 2 marks.

(ii) In Spain, people are prevented from coming in by using radar systems and fences and the like to make unwanted people feel unwanted. Tough control systems include higher taxes with little pay for these immigrants and giving priority to local people. Treaties can be signed between countries to prevent people coming in.

> *e* This is an answer to a different question. The command word is 'why' and not 'how'. No marks are awarded.

(c) Climate, ocean and mountains have an influence. The voluntary migration of people from places like Nigeria, due to economic depression, to Spain is prevented by the Sahara Desert (500 km and 14 days travelling). The Atlas Mountains also see great weather extremes and food shortages. In Morocco, barbed wire fences prevent immigrants, as does the sea between Morocco and Malaga in Spain. All these factors mean fewer people try the voyage and those who survive it are the fittest. People do not travel in extremely large groups because they may hinder each other; also very small groups may mean people are left helpless. Hence, the physical factors influence the volume of people.

> *e* The ideas presented here are appropriate, but they lack any sort of organisation. Only one side of the story is presented and it strays into irrelevance. Remember that you cannot lose marks for irrelevance, but you do waste time. A balanced answer would include physical factors attracting migrants, such as a better climate in the receiving country. This answer is at Level 2 and worth 3 out of 6 marks.

Answer to question 5: candidate B

(a) (i) Forced migration is when migrants have no choice but to travel to a new destination. This may be due to eviction or deportation from a country.

> *e* This response could have been strengthened by a physical/human example. However, it answers the question and is therefore awarded 2 marks.

(ii) (1) Forced migrants can be of any age, but, if due to war, they may only include women and children. Voluntary migration may involve entire families or just male guest workers, giving an unbalanced population pyramid. Some forced migrations, like the slave trade, were age-selective, taking only the young.

e This is a good response. Although it is lacking in detail, it recognises variation and is worth full marks.

(2) Forced migration is more likely to be as close to the country of origin as is safe, especially with refugees, as it is usually temporary. This is clearly not the case with the slave trade, which involved journeys of many hundreds of kilometres. Voluntary migration is usually driven by economics and thus likely to be to a developed country and involve individuals or groups with high aspirations.

e This is a good answer, for full marks. Once again, the approach is theoretical rather than through examples, which is perfectly acceptable unless the question specifically asks for examples.

(b) (i) Governments may encourage internal migration in the first stage of an ethnic cleansing programme, or if they are trying to move people into new areas which they want to grow in economic strength. They may also want to use land for a development and therefore try to remove the inhabitants, as happened in the US national parks. New towns might be built, which lead to shifts in the population.

e There are three strands here: ethnic cleansing, land development and new towns. They are all legitimate and motives are touched on in the first two cases. The response is awarded full marks.

(ii) International migration can have a very negative effect if the people who leave are the most qualified and most likely to contribute to economic growth. This is known as a brain drain and happened in Britain 20 years ago. These days it is happening in some Asian countries, like India, from where electronics experts are being recruited to work in Silicon Valley in California for companies like Hewlett-Packard.

e There is only one idea here, but it is both well handled and illustrated, for full marks.

(c) In Africa, many migrants are eager to reach Europe or warm areas in west Africa, such as Ghana. Although many begin the journey to the Mediterranean, many perish on the way, with a lack of water and food and having to journey across the Sahara Desert and the Atlas Mountains. The Mediterranean crossing is the final hurdle, with many drowning when they become exhausted through swimming or when their boats capsize and sink.

e This is a Level 2 response, worth 3 marks. It is a little clumsy in its approach, rather brief, and volume is not addressed directly. The physical factors addressed are all obstacles rather than positive or negative push and pull factors.

Population movements (II)

Study the diagram below, which is the Zelinsky model of migration, and answer the questions that follow.

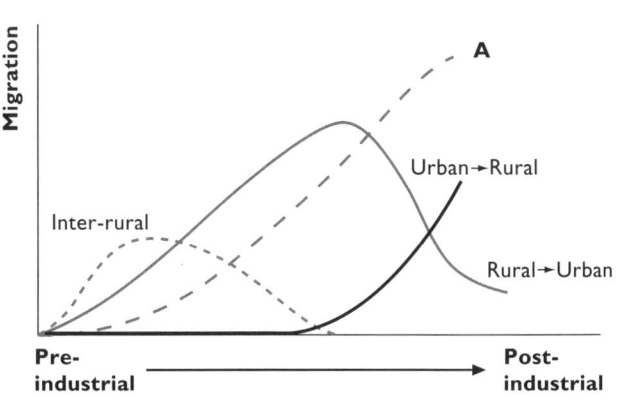

(a) (i) Define the term 'internal migration'. (1 mark)
With reference to the diagram,
(ii) describe the trends in migration from rural areas. (3 marks)
(iii) what kind of migration is shown on line A? (1 mark)
(b) In what ways might the decision to migrate be influenced by:
(i) social factors? (4 marks)
(ii) economic factors? (5 marks)
(c) With reference to located examples, outline the ways in which a large volume of migration can have both positive and negative effects. (6 marks)

Total: 20 marks

■ ■ ■

Answer to question 6: candidate A

(a) (i) This is movement internally.

e This is a simple repetition of the question. No mark is awarded.

(ii) This starts off with a few people moving, then by the middle period it is the most significant of all of the groups and then it declines to be the least important by the end of the time period.

e Only rural–urban movement is considered here, but some comparative comment is offered (most, least), which takes it to 2 out of 3 marks.

(iii) International migration

e This is wrong. The correct answer is **urban–urban migration**.

(b) (i) Many migrants move because of social reasons. They think that their society is intolerant and will not let them practise their religious beliefs or their culture.

83

They come to countries like Britain, where there is less prejudice and where they can live with other people who share their lifestyles and habits.

e This is slightly confused, but makes a valid point about religious beliefs (although it is not located). The point is not developed (what religion? where do they come from?), so 1 mark is awarded.

(ii) Most migrants are migrant workers, like the Turks who went to Germany after the war. They left poor villages in their country to get jobs in German factories, making cars and other manufactured goods. The wages they received were higher than anything that they could get at home, and the housing and health facilities were much better. Sometimes, however, they were racially abused.

e The only factor here is the differential wages in the two countries. Much of the rest strays into social factors (housing and health), which would have been useful in the previous answer but, sadly, cannot be credited here. Since the villages at home were 'poor', some idea is grasped of push factors as well as the wages as a pull factor. The answer is awarded 2 out of 5 marks.

(c) If too many migrants arrive in a country, there will be a shortage of jobs. So in Germany, after a while, there was an economic recession and lots of jobs disappeared. As a result, both Germans and Turks lost jobs, leading to resentment and racial prejudice amongst the Germans. Many Turks were born in Germany and so Germany had become their home and the Germans could not force them to go back. So although they benefited the German economy when the jobs were there, they became a cost when the jobs went.

e This is very simplistic, with one idea rather weakly located. There is a recognition of the positive impact as well as the negative impact, but no development beyond a simple statement. It is a Level 1 response and is awarded 2 marks.

Answer to question 6: candidate B

(a) (i) This is movement within a country or region.

e This is fine, for 1 mark.

(ii) At a low level of economic development, there is a large amount of migration from one rural area to another, for example in a subsistence economy. However, as development continues, this gradually decreases. During the development, there is a gradual, but significant, increase in rural-to-urban movement, with a large quantity of migrants. However, this decreases rapidly at the end of the model to a low, but constant, level.

e This takes a while to get going, but ultimately covers all of the points of both lines — rural–rural and rural–urban.

(iii) Movement from urban to urban areas.

Edexcel (A) Unit 2

🅔 This is correct, for 1 mark.

(b) (i) There may be a large volume of migration between countries that share a common border. For example, 8 million Mexicans live in the USA. A large number of these have moved to join members of their family who already live in the sunbelt states. Some may be moving to improve their 'lifestyle'.

🅔 This is a bit thin. The only clearly social factor identified is rejoining the family. The material about migration to the sunbelt from Mexico is accurate, but not useful here. One developed idea gets 2 marks.

(ii) People are likely to be attracted to areas that have relatively high wage levels. For example, wages are seven times higher in the USA than in Mexico, and good employment opportunities may be attractive, especially to those who live in areas with a surplus of labour, for example in response to mechanisation of agriculture, which has taken place in Mexico. The jobs might be more secure and more chances of promotion might be on offer. Also, areas of rapid economic growth are attractive due to the characteristics of a booming economy, relative to an economy in economic decline and deprivation.

🅔 There is quite a lot here, but the last two lines are rather wasted because the candidate does not specify what these 'characteristics' might be. Elsewhere, a lot is covered in outline, with useful data on comparative wage rates. Push and pull factors are addressed. The response is awarded 4 out of 5 marks.

(c) A large number of migrants may provide the labour force for a growing and developing economy. For example, sunbelt regions are today nearly dependent on Mexican workers in sweat shops and for semi-skilled manual and service work. They once worked in the orange groves. Also, the exported surplus labour from Europe to the USA triggered and supported rapid industrialisation.

In addition, migrants may increase spending in the local economy which sets up a multiplier effect. Some argue that huge levels of migration can place a strain on the welfare state and health services, but there is little actual evidence of this. Migrants send money back home in the form of remittances. There are social and cultural impacts, both on receiving countries and donor countries. The USA has been made by migration and, on a smaller scale, the impact of black athletes is obvious in Britain.

🅔 The candidate uses the same case study here and covers a number of effects, both positive and negative. The quality of language is impressive and although there is a lack of locational detail, it is worth full marks.